The Making of a Hardrock Miner

Miner drilling with a stoper — similar to a jackleg —
to begin driving a vertical raise. (Gardner-Denver Co.)

the making of a

HARDROCK MINER

by Stephen M. Voynick

An account of the experiences of a worker in copper, molybdenum and uranium mines in the West.

THE MAKING OF A HARDROCK MINER

Printed and bound in the United States of America

Library of Congress Catalog Card No. 78-56574

ISBN 0-8310-7116-8

Second Printing 1997

Printed in the USA by

*M*ORRIS
PUBLISHING

3212 E. Hwy 30
Kearney, NE 68847
800-650-7888

DEDICATION

They who always labour can have no
true judgement . . . These are amongst
the effects of unremitted labour, when
men exhaust their attention, burn out
their candles, and are left in the dark.
— *Edmund Burke*

CONTENTS

FOREWORD

Prior to 1968, I had lived most of my life in the white-collar world surrounding New York City. Like the majority of Americans, my understanding of underground mining was quite limited; I considered it to be simply a dangerous occupation pursued by strange men in remote places, something beyond the realm of conventional experience. My total concept may have been embodied in a single photograph of a man in a hard hat with a miner's lamp, coveralls, and a safety belt. That man, impersonal, yet congenial if he could have talked, was operating a rock drill in a nameless underground cavern. Accompanying the photograph was a dry text which further insured his anonymity by telling only cold statistics of the mineral industry.

Quite suddenly in 1970, I found myself, for want of a paycheck, wearing the hard hat and lamp of the hardrock miner and riding a cage into the dark, cold depths of a Colorado mountain. That descent marked the beginning of an odyssey that took me to four mines in three different western states, from the Colorado Rockies to the Arizona desert to the high plains of Wyoming. Along the way, I learned firsthand of that little-known and least publicized facet of underground mining—the human element—and how men, for a paycheck on Friday, can adapt to and endure a hostile world, dark and dangerous.

This book is an account of my experiences in the hardrock mines of the Rocky Mountain West. It is not intended to be a technical account and, therefore, contains no profits and losses, ore reserve estimates, annual productions, or other statistics and figures which compose so much of the popular image of the mining industry. Rather, it is an account of the mines themselves, what it is like to work in them, and mostly of the men who do.

Stephen M. Voynick

Leadville, Colorado
August, 1977

CHAPTER I

CLIMAX – THE NEW HAND

January 1st, 1970. Denver, Colorado.

Like the cries of a wounded duck, the last strained squawks echoed from the noisemakers and the last bits of gaily colored confetti drifted down to litter the beer-splattered floors. The new year was already several hours old, closing time had come and gone, and soon a gray dawn would rise over the eastern plains, bringing with it a sad finality to the revelry and an abrupt return to reality. The great annual escape that is New Year's Eve had run its course and the licenses had expired which permitted supposedly sane people to act like idiots, to say and do things they would regret for the next twelve months, and to judge their level of enjoyment and fulfillment by the degree of physical discomfort to be endured the next day. Die-hards clung to the hope that the sun would not rise, that the party would go on forever, that the joyous artificiality would prevail. Deep down inside they knew too well that life would only grind on, taking them with it, and that January 1970 would amount to little more than a rote continuation of December 1969.

This is a resignation experienced particularly by those shackled to the endless drudgery of long-term careers, the twenty- and thirty-year pension specials in which, if you keep your nose clean, you'll pretty well know what you'll be doing and where you'll be doing it fifteen years from the day you wonder about it. In refreshing contrast to that lackluster static security, there is a positive excitement and fascination in not knowing the whats and wheres of next year, or even next month. It resembles the anticipation that comes with unwrapping a present, but on a far grander scale, for it is the curiosity of life itself. The only essence that can transform life from a plodding, animal existence to a worthwhile experience is said to be love, but too often love becomes tempered and tamed, compromised by social demands. In the end, is relegated to the position of buttress, making those long-term common jobs bearable.

The uncertainty that accompanies such independence, however, is not something always readily managed. A poll of winos and prison inmates would confirm this, revealing that the majority had bounced around in various colorful pursuits until at last that exciting uncertainty had soured to disillusionment and discouragement. In such cases, a long term career, however banal, might have provided the balance and stability that was lacking. The most interesting people I have ever met were the ones who could meet these random situations, however foreign to their backgrounds, as they arose, and who assimilated easily to become an integral part of new people and new places, and yet still maintained a detached perspective that enabled them to see the woods despite the trees. Society, in its self-righteous judgement, labels as drifters, misfits, and bums these fortunate few for whom life is an honest adventure.

Sometimes it is surprising how far one may go on an odd job here and an odd job there. Over the past year, a string of such jobs, including road construction, industrial photographer, commercial salmon fisherman, and, very briefly, cop, had paid the way on a trip from coast to coast and from Mexico to Alaska. Steered only by the winds of chance, this footloose odyssey had painted a portrait of sorts of the country, sketched and colored by many people, many places, and many experiences.

Experience may be academically defined as observations and practices resulting in or tending towards knowledge, a broad and general definition which quite accurately characterized the nature of my recent experiences as just that—broad and general. They were, of course, not the kind of experiences one might recount with a sense of professional accomplishment on an employment application. Personnel managers, dreary men in white shirts and drab ties, our contemporary dealers in flesh, seek another kind of experience, one which might be defined as an enhanced comprehension and efficiency applicable to a specific field. The qualities they seek are based upon the straight and narrow, the adherence to social guidelines, the willingness to accept human, error-prone authority as the final word, and upon some previously demonstrated corporate allegiance. If present, such attributes should be reflected to a greater or lesser degree in the record of past employment which the applicant has listed truthfully and accurately in reverse chronological order beginning with the last job first. A floor sweeper, for example, should only have left his last job to seize the opportunity to sweep bigger and better floors elsewhere. The ideal prospect for this position of floor sweeper would have broken in on concrete warehouse floors and progressed to the polished hardwood of executive suites with nary a dalliance in between. Personnel interviewers, confronted by such a nervously smiling prospect, feel their cold hearts beating faster, realizing here indeed is fodder for the company.

In any odd-job situation, the time arrives, sooner or later, when something a bit more substantial is required to fill the empty coffers, and that time had reared its ugly head and smiled at me on a cold January morning in Denver, Colorado. I needed a job, a real job, but I was poignantly aware that an honest listing of my past recent jobs, or rather job experiences, would bring simultaneous laughter and teardrops to the face of even a hardened personnel manager.

The mythical job I envisioned would have to offer quick hiring with a yes or a no on the spot, pay enough to save something, and, for the sake of interest, involve something different. Such a job certainly exists, I told myself, as I prepared for the methodical, alphabetical search through the voluminous employment listings of the Sunday *Denver Post*. I had hardly begun when my eye was drawn to a large block ad with the caricature of a miner in hard hat and lamp. The thoughtful artist had even drawn in a beam of light emanating from the lamp, in which appeared the words "Shine A Little Light Into Your Future." Beneath that, in bold print, was "Hiring Immediately—No Experience Necessary," a phrase tailored to the first of my requirements. The ad went on, wasting no words and getting right to business, stating the pay and listing the benefits that went with the job of "Underground Miner." All the prospective employee had to do was show up and pass a "rigid physical examination" at the Climax Molybdenum Company, Climax, Colorado.

In January, the drive to Climax from anywhere is an ordeal, particularly in those days when the Eisenhower Tunnel was only a funding question and tiny caravans of cars and trucks, each led by an enormous blue-lighted snowplow, crept slowly over the Continental Divide at 12,000-foot Loveland Pass. After three hours of icy mountain driving, the sign that materialized from the whiteness of the snowbanks and ground blizzards announced arrival at Fremont Pass, elevation 11,318 feet. Just beyond that, a second sign pointed out the gate of the Climax Molybdenum Company, same elevation. I was beginning to realize why they advertised to get people up here.

True to its advertised word, the Climax Molybdenum Company wanted bodies and cared not what they had done and where they had been. Their main concern seemed to be with the condition of the backs on those bodies. Blizzard notwithstanding, the employment office was doing a good business, methodically steering men between interviewers, clerks, and doctors, a polished, efficient routine perfected from repetition. After signing a few forms, answering a few questions, and urinating in a bottle, and with noted absence of congratulatory ceremony, I found myself hired as a "Miner's Helper" and instructed to report Monday morning for orientation.

When Monday arrived, I learned that the orientation was hardly an exclusive courtesy, but something to be shared with twenty-six others who wanted to shine a little light into their futures also. Bits of subdued conversation buzzed around the spacious room while the bodies waited to be oriented, mostly rationalizations of why the particular speaker was on Fremont Pass in January seeking work in an underground mine—tragic tales of shattered loves, unreasonable parole officers, lay-offs, hungry children, and no money for next semester. The murmur faded out as a thin, wiry man with a gray crewcut and a face full of wrinkles walked to the front. With the exaggerated military bearing of a master sergeant about to address the raw recruits, he cleared his throat, accenting the silence, and let his eyes roam over the twenty-seven waiting faces. After introducing himself, he began in a low, steady voice.

"Anybody here done any minin'?"

A nasal twang from the back said something about helping his old man dig a seam of coal out of the side of somebody's hill.

"I mean *real* minin'. Underground."

Silence.

"That's what I thought." With the point effectively made as to who would talk and who would listen, the bright blue eyes that peered from the forest of wrinkles lost some of their formality and the voice softened a bit. "Then everything I'm going to say to you will be new, so listen up. If you got questions, ask 'em."

Chairs creaked as bodies shifted and got comfortable.

"First thing, remember, nobody led you men up here by the hand. You're here because you think you can make more money here than you can anyplace else. Anybody come for a different reason?"

No comment from the lovers and parole jumpers.

"Good. Then we got no liars here."

That brought a subdued chuckle and broke the ice. This wasn't the first welcome speech the old man had given.

"Men, underground minin' is a little different than anything else you ever seen. Believe me. I've been here twenty-seven years and I seen a lot of men come and go. Some didn't last their first shift, others are foremen now. It's all what you make of it. If you come here to play around and have a good time, you're gonna get hurt and hurt bad. Maybe killed. Worse yet, you might get somebody else hurt, a good man, maybe. There's a thousand ways to get hurt down there and you gotta be watchin' for every one of 'em, all the time. Forget where you are for one minute, one second, and they'll be haulin' your ass out of there in a basket."

The faces of the twenty-seven "miner's helpers," not one of whom had ever set foot underground, could have been cast in stone. Not a single expression revealed even a hint of doubt, of concern, of the soul searching that begins with the realization that this job might not be exactly what they had in mind.

"Now then, we had a man get hurt this past Friday. He got hurt because he didn't pay attention to one of the things that I'll be tellin' you. His mistake was pickin' the wrong side of the drift to let a muck train go by. It derailed on a curve and pinned him in against the rib. Took near an hour to get him out. Crushed chest, ribs, back, everything. I don't mean bruised, I mean crushed." He emphasized the "crushed" slowly, through clenched teeth, and theatrically provided a graphic description by forcefully compressing and grinding the palms of his hands.

I thought I could feel my chest tightening. Nobody really knew precisely what a muck train or a drift was, but there was no doubt at all about the crushed chest. Everyone else must have felt the same thing, too, for the blank faces came to life and twitched, eyebrows lifted, chairs creaked, and a pencil clattered onto the floor. Heads turned just enough to exchange brief glances in the silent masculine manner of seeking reassurance. There was a weak question about the injured miner's present condition.

"Right now, I got no idea." The answer came back with measured disdain and a shrug of nonchalance. "They had him here in the hospital for a little while, then took him to Denver. If he does come out of it, he won't be worth a damn for a long time."

On Tuesday, only nineteen Miner's Helpers showed up to continue with the orientation. The old man, Jim Wizen, smiled and took up on a different note.

"Okay, we can get started seriously today. If you look around, you'll see that eight of your buddies are gone. Decided this wasn't for them. That's exactly why we didn't issue you any gear yesterday, just saves the trouble of turnin' it back in so quick."

The gear we did draw that morning filled both arms and consisted of a hard hat, D-ring safety belt, high-top rubber boots with steel toes, respirator, safety glasses, ear plugs, and a couple pairs of gloves. All of it was new and stiff, and in keeping with the Army tradition, nothing fit right. The other item we received was brass. I picked up the shiny, half-dollar-sized brass disc and examined it. Deeply stamped across the face was the number 11982, my Climax employee number, which told that 11,981 nameless souls had worked underground at Climax before I got the bright idea.

"The brass system here works like this," Wizen explained. "The only time you carry your brass with you is when you're in the hole. Any other time, your shift boss will have it hanging on a rack behind his time window. When you go on shift, you pick it up and carry it with you. Comin' off shift, you turn it back in. No exceptions. This is the only way we can positively account for everybody in the hole. If something should happen and you can't get out, maybe you're laying hurt somewhere and can't move, this is the only way we're going to know about it. So again, when you come off shift, don't forget to brass out. If you do forget, somebody's gonna be down there looking for you, and if you're up here takin' a hot shower or sittin' on your ass in Leadville drinkin' a beer, well he ain't gonna like it."

On Wednesday, we were issued numbered lamps and baskets in "the dry," the change and shower facility which is similar to a large athletic locker room in both appearance and odor. The wire baskets in which the miners stowed their gear were suspended from the roof supports and run up and down on individual chains. To accommodate the large Climax shifts, the Storke Level dry was nearly One hundred feet square and lined with several spacious shower rooms the Army would have been proud of.

Wizen handed out diagrams to two of the mine levels, the Storke and the 600, that looked like the maze puzzles kids play with to see whether they can get the steel ball back to "start." Each tunnel, or drift, was assigned a four-digit number and direction, and Wizen told us how easy it was to find, say, Sixty-two Ten South. "Then, when you get to one-switch, you take the south hangin' wall, go up six drifts. That's all there is to it, See?"

Everyone nodded and grunted in the affirmative. No one knew what he was talking about.

"Okay," he said happily, obviously pleased with his own apparent instructional abilities or our exceptional comprehension. "Keep those little maps in your pockets and you'll never get lost. Now, let's get into our gear and we'll take a look at the underground."

We put on the boots, hard hats, strapped the lamp batteries onto our belts, clipped the lamps to the hard hats, hung respirators around our necks, put on the glasses, and followed Wizen out to the Storke Level portal. A large network of railway tracks became reduced to a single track at the concrete entranceway. The portal itself was covered by a metal snow roof to keep the tracks and switches clear of the snow, which, at this elevation, could fall at any month of the year. After a final head count, Wizen led us through the portal into the adit, the long entrance tunnel leading to the interior mine level. Nineteen Miner's Helpers followed, tripping and stumbling, cursing, trying to adjust lamp beams under ill-fitting hard hats. Two miners, waiting for a lift on an inbound train, gave Wizen a wave for recognition, glanced back at his trailing carnival troupe, looked at each other and shook their heads.

When we had gotten about two hundred yards into the adit, Wizen called us together. "Okay, now listen," his voice echoed off the rock walls. This track we're standing on is the only one in or out of the Storke Level, so every single train uses it. One-switch is still another half mile further, so we get a lot of traffic here. But we got a siren warning system that'll tell you just what the traffic is doing, so all you got to do is listen to find out what's happnin'. One blast means a train is comin' out, and two means you got one inbound. When you hear either, start looking for a place off in the rib, one of them manholes. That's what they were made for and they're the safest place to wait out a passing train."

As if on cue, a long, single, haunting wail of the siren echoed down the adit and Wizen herded us into a manhole. A light appeared in the distant gloom and a deep, growing rumble preceded the approach of the muck train. A thirty-ton motor, a heavy, brutal, filthy looking beast, bathed in the alternate flashes of blue and white light from the arcing trolley runner overhead, lumbered by with the oddly human and vulnerable face of the motorman peering from a grimy window. Loaded muck cars, twenty of them, rumbled by lurching precariously from side to side. I thought of the guy with the crushed chest.

When the adit was quiet again, Wizen said, "The only other signal is a steady wail. That means all traffic has to clear the adit because they're haulin' out an injured man."

Now there were only twelve Miner's Helpers. The remainder of the week was spent divided between underground familiarization and classroom lectures in which safety was nearly the sole topic. There were literally hundreds of safety regulations that applied to underground mining and Wizen dutifully tried to explain each. This was more an exercise in rhetoric than anything else, for the language of mines and miners is unique unto itself and, for the uninitiated, is very confusing. Accurate descriptions of underground operations and working mines are, for that reason, rarely conveyed to persons who have not themselves worked underground. The miners' vocabulary has been almost singularly responsible for isolating their profession from the more common and widely understood currents of labor experience. Listening to Wizen read, translating as best he could the long list of regulations, served only to create a general awareness of the myriad of ways to get hurt or killed underground. Full appreciation and respect for the hazards of underground mining, I would learn, are things which develop only with time and experience.

During the brief forays underground, the basic elements of the new language came into use. What had always been a tunnel was now a drift, the sides, or walls, of those drifts were ribs, and the overhead was the back. "Going underground," for the class of neophyte miners, meant little more than that; walking along haulage drifts, keeping clear of passing trains, and, occasionally, cleaning out a few railway switches with mucksticks, the miners' term for the lowly shovel.

Wizen delivered many underground lectures, the longest and most detailed of which dealt with the phenomenon of falling rock, pointing out repeatedly that this was the single largest cause of underground injury. Walking slowly along the dark drifts, he studied the rock over our heads, pointing out certain features he deemed worthy of our attention.

"Now see that slab there?" he asked, casting the beam of his cap lamp over a suspiciously protruding rock about the size of a garbage can. "She's cracked clear 'round. You can see it. Wouldn't take much to bring 'er out of there. Slab that big could kill a man if it hits 'im right."

The offending killer slab, now bathed in the weaving lights of a dozen cap lamps, became the object of close collective scrutiny.

"How often them stones fall off the ceiling?"

Wizen smiled wanly and paused a second before answering. "You mean, how often do them slabs come out of the back?" The old man had the patience of Job.

Another pause while his student contemplated the semantic correction. "Yeah."

"Now that's something that's up to you. Up to each one of you. The very first thing you'll do when you get to your working places at the start of every shift is bar that loose crap down. You got scaling bars all over this mine and you're expected to use 'em. Anything you bring down can't fall on you later, right?"

A chorus of umms. With all heads turned upwards in awe, there was enough light on that rock to make it grow.

Producing a scaling bar, an eight-foot-long steel bar with a flat, angled tip, Wizen ordered, "Okay, now everybody get out of the way, get back where you can watch this."

Gingerly poking the celebrated hanging rock which, by now, seemed to have grown in size, Wizen inched closer, probing for leverage. Getting a bite with the bar, he pressed downward and the rock moved a few inches with a heavy, groaning sound. The old man, who doubtlessly had barred down thousands of these things in his long career underground, moved still closer to get a better bite and execute the *coup de grâce*. When the bar had barely touched the ruck, it groaned again and fell out with astounding speed, smashing into the ground about two feet in front of Wizen's toes with a sickening thud. Not quite expecting it, Wizen lurched backwards, catching his feet on the rail and falling on his back with a second mighty thud. Had he stood his ground, it would have been a truly magnificent performance.

But such glory was not to be. For a long minute, the old man lay in the drift, his agony and embarrassment spotlighted by the frozen beams of a dozen cap lamps. When it had been ascertained that it was his pride that had absorbed most of the impact, he wearily shrugged off the offers of assistance and climbed slowly to his feet in the awkward silence.

Wizen looked calmly and thoughtfully at the rock that had fought him to a draw and, with a wry grin, quietly mumbled a phrase I would hear again. "Gettin' too old for this shit."

At the end of the orientation period, I really didn't know much more than when I hired on a week earlier. I had not worked with a miner, had not seen a working heading, about which we had heard so many tales, and had not ventured underground much farther than the Storke portal. But all that was to change soon, for having been sufficiently oriented, I was assigned to a mine crew on the 600 level.

"You'll be working for Kenny Reasoner," Wizen told me, hand on my shoulder. "He'll be your shift boss. Good man to work for. Report to him day shift Monday in the Storke dry. And good luck."

That was the last I ever saw of him.

The Climax Molybdenum Company mine and mill are perched near the top of the Continental Divide in central Colorado's Ten Mile Range, a long line of 12,000-foot granite peaks. The mine elevation of 11,300-feet makes Climax one of the highest underground mines in the United States, and places it right at timberline, where the dark green carpet of pine and spruce below gives way to the barren rocky crags and perpetual snow drifts that cap the Rockies. Snow can and does fall every month of the year here and the average snowfall, measured by the snow accumulating from October to June, is about two hundred inches. Bad winters dump three hundred inches.

Climax, as most passing travelers see it from Colorado highway 91 on Fremont Pass, disregarding its unusual location, would pass for a large industrial plant anywhere in the country. The sprawling complex of gray, corrugated steel buildings, company streets, railroad tracks, power lines, and conveyors is suggestive of an operation that should employ more than two thousand workers. Directly behind the mill buildings rises Bartlett Mountain, or rather what is left of it. The entire western shoulder of the mountain has been reduced to a gaping "glory hole" of immense proportions, a garish, yellow cavity similar in shape to an immense volcanic cone nearly a half mile across. All passing travelers stare at the glory hole, but few realize, or can comprehend, that it was created through the efforts of a half century of underground mining. Three main haulage levels, with more than thirty miles of underground railroad and one hundred miles of underground workings, extend beneath Bartlett's massive, internal ore body. Creation of this underground world required the removal of over 400 million tons of muck, or broken rock. Such a quantity of rock, or of anything else, remains only a statistic, a meaningless number on the scale of the national debt, beyond the realm of true comprehension. That 400 million tons of dynamited rock was pulled from the guts of Bartlett Mountain by trains of twenty muck cars each, each car laden with ten tons of rock. Slowly, ever so slowly, the surface ground was allowed to cave into the mined out area, eventually resulting in the gross transfiguration of the mountain itself. An ecologist's nightmare, the Climax glory hole must nevertheless be considered one of the unsung monuments to man's patience and determination.

The object of all this effort is molybdenite ore, a white rock shot through with dark streaks, a ton of which, after milling, yields only four pounds of molybdenum sulfide concentrate. Further treatment produces the metal and its derivatives, which have found applications in lubricants and space-age steel alloys. Along with the "moly" iron sulfide as pyrite is also recovered, as are lesser amounts of tin and tungsten.

The Storke dry had a row of twelve time windows, very much like the pari-mutuel windows at a race track. Each crew, whether mine, track, form, cement, repair, or muck, has its own window where the shift boss hands out orders and brass at the start of a shift. About a half hour before the day shift's 7:00 a.m. brass-in time, the miners drift in from the change rooms, remove their lamps from the charging racks, and form little groups of three or four, sitting and lying on anything that is remotely comfortable.

Ragged coats, trousers, and shirts are covered with dirt, stains, and grease, and have been worn and washed into a drab uniformity. Hard hats, caked with months' accumulations of dirt, are worn cocked at every conceivable angle. Safety belts, supple from wear, are slung low on the hips. Lamp batteries, wrenches, and tools of all types dangle from the belts. The miners walk slowly from the lamp racks, their tools making a curious muffled jingling sound, deciding upon which group they will grace with their conversation. Selecting one, a miner elbows in with a nod or grunt of greeting, then invariably reaches into his shirt pocket and withdraws a can of snuff. Tapping the top twice for good luck, he opens it and takes a pinch, placing the moist, finely ground tobacco deftly behind the lower lip, silently offers the can around, caps it, and stuffs it back in the shirt pocket.

The conversations before a shift are generally subdued and quiet, with each man finding nothing to be particularly happy about. The eight-hour shift coming up is not looked forward to and the mood is clearly one of resignation to just get it over with. Expressions are serious, bland, or meaningless, and the few smiles seem forced or of a sarcastic variety.

The degree to which a new man stands out in this crowd is astounding. From the moment he makes his appearance, he becomes the subject of cool, appraising stares from the miners. Although they never stop their conversations, they are plainly amused by the spectacle of that new man, from his clean boots to the stiff new safety belt and gloves that have never seen use, to the clothes that have never seen real dirt, and up to the shiny hard hat. It is that hard hat, more than any other single thing that clearly announces a new man about to masquerade as an underground miner. The experienced miners seem to blend into their scratched, battered hard hats and wear them at angles that make the head and the hat, even with the protruding lamp, appear to be a single homogenous unit. On a new hand, the lamp-hat combination rests precariously on the head like a fat barnyard chicken trying to roost in a tree. It's there, but it doesn't want to be.

I adjusted that damned hard hat for the tenth time in the five minutes I had had it on, futilely trying to compensate for the forward weight of the lamp, and walked up to a circle of four miners. "Excuse me," I said politely, as if about to inquire directions to the office of the Dean of English. As soon as the words were out, I wished I had never said them.

Four heads turned abruptly in unison, and cool, curious eyes peered out from beneath the hat brims. If there were one hundred men on that floor, I realized belatedly that not one of them would have begun a sentence with such a superfluous nicety as "excuse me."

"Ahh, can you tell me where I can find Kenny Reasoner?"

The ensuing silence was born not of conscious discourtesy, but of honest surprise and curiosity, as if a talking monkey had asked the question. Feeling increasingly uncomfortable and foolish, I reached up again to adjust my hard hat, probably to make sure it wasn't on backwards. The miners exchanged amused glances, ascertaining silently among themselves whether it was all right to tell this guy where Reasoner was. Apparently it was, or else they just wanted to get rid of me, for the answer came as a slow, pointing gesture directing me toward the end time window. The obliging finger was missing the last knuckle.

The face behind the window was cast downward, engrossed in the handwritten pages of a small time book. As I walked up, the pen stopped moving and only the eyes looked up to assess the new arrival.

"You Kenny Reasoner?" I asked.

Neither the angular face nor the gray, placid eyes reflected any expression whatever. Keeping his eyes fixed on mine, he opened his hand slightly and extended it in my direction. I filled it with the assignment slip Wizen had given me. Well, this must be Reasoner, I guessed, as he straightened to his full height, several inches over me. He held the slip of paper in his left hand and read it while tapping the pen in his right hand slowly and rhythmically on the countertop.

"You got any brass?" he finally asked.

I produced the brass disc and gave it to him. Number 11982. I imagined his number to be something like 6 or 7. Reasoner now turned his attention back to the time book as if I had vanished into thin air. I stood there waiting for a moment as if something would happen. When it didn't, I asked the natural question. "Well, what do you want me to do?"

Again the eyes lifted, this time with a slight indication of annoyance. The left hand opened, palm down, in a soothing, patting motion. Hang loose, don't get excited.

The floor was full of miners now, all gathered against the far wall, leaving a fifteen foot void in front of the time windows in which I was the sole occupant. I found out why. A distant rumble grew into a thunder which actually shook the dry. The double swinging doors at the top of the stairs suddenly burst open, slamming violently against the concrete walls. A surge of humanity, a gray-brown flowing mass with grimy human faces beneath the cyclops-like lamps, charged down the stairs, jumping, pushing, running, and leaving a trail of mud from a hundred pairs of stomping mine boots. Feverishly wide eyes, brilliantly white against all the dirt, shone with determination and excitement. A cacophony of yelling, scuffling of boots, rattling of tools, clattering of lamps being slammed against racks, and the metallic jingle of brass being thrown into the time windows resounded through the dry. Here and there, an involuntary lurch generated by a good goose would be followed by a chorus of raucous laughter and an obscene scream of promised revenge. I was bumped, pushed, and shoved in a futile effort to fight the tide, but it was over almost as quickly as it had started. A few stragglers drifted down the muddy stairs, but the riot had already shifted to the change and shower rooms. Graveyard shift had come off and it was evident they wanted out.

Day shift moved slowly towards the windows, each crew queuing up in front of their respective shift bosses. I fell in at the end of Reasoner's line. Each man picked up his brass and listened to the quietly spoken instructions. When it came down to me, Reasoner flipped my brass out on the counter and gave me a long, contemplative look that spoke for itself. What the hell am I going to do with this guy? Glancing over his crew of twenty men just starting to move towards the portal, he called out, "Dwayne!"

The miner who returned to the window was thin, sported a black Vandyke beard and the beginnings of a waxed handlebar moustache. He was not much older than I.

"Dwayne, I want you to take the new hand with you," Reasoner said, cocking his head in my direction.

The miner stared for a moment at Reasoner, shot a quick glance of appraisal at me, then turned slowly back to his shift boss. His expression was neutral, but still showed an air of resignation. Why me? Turning back to me, he asked, "What's your name?" with the weakest smile of attempted congeniality I had ever seen.

"Steve."

"I'm Dwayne."

That was the last time he would ever call me Steve and I would ever call him Dwayne.

"Let's go, Pard," he said. I followed him towards the portal.

"Hey, Dwayne," Reasoner called out again. When we had turned back toward the window, he went on. "Keep an eye on him, now. New hand."

That's me. The new hand.

Underground mining, reduced to its simplest terms, means the breaking of rock and its removal. Although many different kinds of crews simultaneously worked underground at Climax, their effort is all in support of the mine crews whose job is the basic breaking of rock. In a cave system of mining, such as is utilized at Climax, an elaborate network of haulage drifts must first be constructed beneath the ore body. Reasoner's crew and one other mine crew was engaged in that development work with two men usually working as a team in each heading. It required at least two men to handle the equipment. Also, for safety reasons, company policy prohibited any man from working alone underground.

The haulage drifts were about twelve feet square on a rock-to-rock measurement. Most were supported by sets of twelve-inch-square timber standing six feet apart, each set consisting of two vertical posts crossed by a horizontal cap, all ten feet long. Cribbing, shorter lengths of smaller timbers, are then stacked on top of the caps to connect each set and to support the overhead rock in the back. The entire structure is secured not with nails, but with wooden wedges which are driven with axes into the appropriate spaces in the timbers. A properly installed set will "take weight" almost immediately, countering the slight inward constricting rock movement which begins as soon as the rock has been blasted, and providing stabilization for the drift.

The cribbing also serves to support any large slabs which may work loose from the back. Lagging, three-inch-thick boards, are laid cap to cap to keep the "small stuff" off the miners' heads.

The miners who work with the rock, against the rock, and literally in the rock, have personified that cold, hard, unyielding material and refer to it as "she." When a new hand listens to miners talking about the condition of the rock in a particular heading, all he hears are lines. like "she ain't breakin' right," "she's gettin' a little ratty now," or if the rock is really bad and there is serious slabbing, "she's plumb bad in there, don't take your eye off 'er cause she'll try to getcha." For all a new hand would know in his stumbling ignorance, there could be a couple of nasty broads slinking around each heading.

If the rock is "good," or particularly stable and not prone to slabbing excessively, timber sets may be unnecessary and rock bolts may provide the desired degree of ground support. The steel bolts, used in four-, six-, and eight-foot lengths, have an expanding head, or shell, to grip the rock. They are inserted into drill holes in the ribs and back. Pneumatic wrenches are used to tighten the bolts against a metal plate, similar to a large washer, flush on the rock. In theory, the angular placing of the bolts in a radiating pattern out from the center of the drift will lock the exposed rock, that which will aerate and tend to slab, into the interior solid rock mass. When rock bolts are used, the drift itself will be blasted out in an arch shape, with the ribs gradually meeting the back in a manner that provides considerable natural support. If the rock is deemed to be "very good," usually by someone who doesn't have to work under it, even rock bolts may be unnecessary and drift development will proceed "bald headed." Generally, Climax relied on timber sets for ground support, and inestimable millions of board feet of timber lined the many miles of underground haulageway.

The work of drift mining takes place at the face, the solid rock which must be broken and removed if the drift is to advance. The last set of timber will extend nearly to the face, providing protection and support at all places except at the face itself. Twelve-foot drifts are relatively large by underground standards and drilling is done with a jumbo, a mobile unit mounting heavy rock drills on movable arms. Jumbo drilling is a fairly safe operation, with both miners sitting on the unit about two sets back from the face controlling the hydraulic arms which position the pneumatic drills. A ten-foot-long drill steel, tipped by a carbide steel rock bit, is attached to each drill and the jumbo is moved forward until the bits nearly touch the rock face.

The stillness of the drift is utterly shattered when the drills come to life. With both a rapid rotary and a vicious hammering motion imparted to the steels, the carbide bits attack the solid rock. The resulting noise and vibration, resonating within the narrow confines of the drift, is deafening. A large jackhammer or pavement breaker pounding away on the surface is mild by comparison. All drilling is done wet, with a stream of water forced through hollow cores of the steels to the bits. Mixing with the finely pulverized rock, the water forms a mud which acts as a grinding compound and also flushes out the ever-deepening drill holes. Most importantly from the miners' standpoint, it prevents the buildup of a suspension of rock dust in the air. In the early days of mining when mechanical drills were first introduced, drilling was done dry out of ignorance, and the results were tragic. Inhaling quantities of the fine silica dust induced silicosis, a pneumoconiosis of the lungs which petrified the delicate lung tissues and caused the slow, debilitating death of many a miner.

The drilling operation produces a pea-soup fog which may limit visibility to ten feet, barely enough to see the face, even with the help of the jumbo's two spotlights. The combined effect of the pounding vibration and concussion, the shaking beams of yellow light stabbing into the thick mist. The glimpses of rock jarred loose and falling from the back, the throbbing maze of high pressure air and hydraulic hoses, and the deafening roar emanating from the two thundering rock drills create a scene straight from the cellars of hell. Barring mechanical breakdowns and hung steel, two miners may drill out the face in about an hour.

The resulting pattern of nearly fifty drill holes, each ten feet deep and two inches in diameter, is then loaded with six cases—three hundred pounds—of gelatin dynamite. A time delay electric detonating cap is inserted into a stick of dynamite, or powder, to form a primer, which is the first stick placed into each hole. That stick is followed by four or five others and the entire string is rammed home with a wooden loading stick. Only one or two holes are left empty, both in the burn, the tight center group of five holes. The electric delays, numbered 0 through 10, will detonate over a period of about eight seconds. The 0 delays detonate instantaneously to collapse the empty holes, creating a shattered, hollow core within the face. The remaining delays, placed so each progressive detonation works against that enlarging core, will bring down the ribs, the back, and finally the lifters, the lowest row of holes.

A loded face resembles a colorful but deadly Christmas tree with red and green lead wires dangling from each hole. These are connected together to form a series circuit, which, in turn, is wired into the main blasting circuit. When all tools and equipment have been moved back from the face, and after all possible approaches to the heading have been guarded to prevent some lost soul from wandering in, and after the time honored "fire in the hole" warning cry has echoed down the drifts, the little red switch on the firing box is thrown and all hell breaks loose. For eight seconds the rhythmical detonations continue, each announced by a sharp crack and backed by a deeper rumbling, a concussion that shakes the rock over your head and the ground under your feet. An experienced miner can pretty well tell from the sound of the blast whether his round "pulled" right and, more importantly, whether a misfire might have occurred.

Dense smoke and fumes will prevent the miners from returning immediately to view the results of their handiwork, but when the ventilation system has cleared the air, if all went as expected, they should return to find a neat muck pile weighing nearly fifty tons. If a partial misfire has taken place, a rare occurrence, telltale lead wires will dangle from the holes that didn't detonate and the ticklish procedure of rewiring and repriming must begin. Misfires in the top holes can leave great pieces of cracked rock weighing tons just looking for an excuse to crash down. Rewiring involves crawling over the top of the fresh muck pile—beyond the protection of the last set—where falling rock can crush a man or cause an undetonated primer to explode in his lap.

After the back has been barred down, the heading is mucked out with a rail-mounted, pneumatically-powered mucking machine. The small front bucket is rammed directly into the muck pile, then violently flung backwards over the top of the machine, a scant foot or two from the operator's face, throwing the muck into a waiting car. While one man operates this snorting, hissing, wildly lurching machine, his partner is occupied on the motor switching muck cars back and forth as they are filled. The heavy muckers are one of the most dangerous pieces of equipment in the headings, with a nasty tendency to suddenly derail, possibly pinning the operator against the rib or crushing his feet against the rail, a discomforting thought to ponder as you dodge the rocks that may be flung sideways from the flashing bucket. Last, but not least, there is always the chance of slamming the steel bucket into a misfired primer hidden and waiting in the muck pile.

After mucking, a new set of timber is installed, rail is advanced to the face and, if necessary, pipe sections are added to the air pressure, water, sump, and overhead ventilation lines that wind through every drift. When this cycle has been completed, the drift will have been advanced ten feet through the 300-million-year-old rock.

If all goes smoothly, this cycle can be completed by two miners in an eight-hour shift, but since Climax does not pay a contract or in-

Underground electric rail haulage. The muck car behind the motor has just been filled from an overhead drawhole in the slusher drift. (*Engineering/Mining Journal*)

centive bonus for work performed beyond a certain point, it is not. Without the contract incentive, the miners do what they can or feel up to, taking their sweet time to bar down the back, change out defective equipment and, most of all, just make damned sure they don't get hurt.

The eight headings assigned to Reasoner's crew were scattered more than a mile apart on two levels. If Reasoner kept moving and encountered no trouble, he would be able to visit each of his headings twice during the course of a shift, which meant his miners worked on their own most of the time and rarely saw each other except during shift changes and lunch.

In a large, sprawling mine like Climax, a new hand's problems are compounded by the fact that every drift looks exactly like every other drift. He has not yet developed even a rudimentary sense of direction or distance underground. Up and down seem to be the only two directions he is reasonably sure of and tales of new hands wandering about lost for the duration of a shift are legendary. I had been working underground not more than a week when Black asked me to take the motor "over to that supply point we were at the other day" and rustle some timber. Certainly I started out in the right direction since there is but one way leading out of a heading. The little five-ton motor clattered down the dark drifts under the flashing arc of the following trolley runner and I stopped several times to throw the rail switches I thought were necessary to reach the supply point. But the supply point never appeared and, after ten minutes of railroading around the drifts alone, I decided to run the motor back to the heading and start over.

That was a sound idea, but I was unable to find my way back. I wound up in a remote section of the level that had been developed years before and was now only rarely used. For well over an hour I threw switches and clickety-clacked about a mile or more through the dark, ominous, identical drifts before the motor suddenly lost power. The overhead trolley line had been terminated, a point I had failed to notice, and the speed had carried the motor well out onto "dead" track. Abandoning the motor, which was now useless, I set off on foot through the inky darkness and heavy silence that was broken only by the incessant drip-drip-drip of water and the beam of my cap lamp. The floor plan of the level was confusing indeed, with several ways to return unknowingly to the place you had just left. Walking along the ties, I amused myself by considering the number of mathematical possibilities there were in trying every drift in every direction. Nearly an hour and a half had passed since I had confidently set out for the supply point when a distant pinpoint of light approached. Reasoner.

"What the hell are you doing down here?"

"Trying to figure out where the hell I am."

"Where's your motor?"

That was a damned good question. I didn't know where the motor was, but I did know that it would take me another hour and a half, with luck, to find it again. "Back there somewhere," I answered, jabbing a thumb over my shoulder. "I ran out of trolley and left it there."

Reasoner shook his head in mild disgust. A fleeting trace of a grin appeared for a brief second. "Just go on back to your heading, your partner's been worrying about you for an hour. I'll take care of the motor."

"Kenny," I said, "I'm not trying to be funny or cause you any trouble, but right now I couldn't find that heading any more than I could find the motor."

He gave me a long look of strained perserverance. "Okay, okay, the hell with the motor. We'll worry about it later. Let's just go to the lunchroom, it's about time anyway." Since I had imagined myself to be lost in a long-forgotten maze of uncharted and abandoned drifts, the walk to the lunchroom proved embarrassingly short.

Lunches were the great social event of the underground. The company authorized a half hour in one of the underground lunch rooms that had been blasted out of a rib and floored, enclosed, and heated to offset the forty-degree chill that pervaded the mine the year around. The back was rock bolted with a heavy wire mesh, allowing the miners to eat in peace without worrying about loose rock. With half the shift over, lunch saw a noticeable lift in the mood of the crew that was often reflected in the loud, raunchy, animated conversations. Doubtlessly, it was the lights, warmth, and human faces in the lunch rooms, which were such a contrast to the first four hours of darkness and isolation, that brought out the ebullient best in almost everyone.

There is an unwritten miners' law that keeps the new hand apart from the social interaction of the crew. He is not asked to join in any conversation and is quite blatantly ignored. Since most of the talk dwells on the mine itself, there is little a new hand can intelligently contribute, and he usually doesn't know what the rest of the crew is laughing and talking about anyway. His lot is to be content sitting at the far corner of the lunch table feigning interest and smiling on cue when the rest of the crew laughs like hell. Acceptance into the crew can be a time-consuming and complex event that begins with an evaluation of the new hand's performance in the headings.

"How's that partner of yours doing?"

"Not too bad. Don't know much but he gets right after it. Had him muckin' the other day and he damn near turned it over, wheels up. Jumped right back on it, though. Give him a little time and he should turn into a pretty good hand."

"Sounds brighter than that ass they gave me."

"Dumb?"

"Dumb? Ain't got the brains to wipe his butt. Sent him after a case of powder this mornin'. Fifteen minutes later he comes back with four sticks. I says 'Look, Pard, if you're gonna bring it four sticks at a time, we'll spend the weekend down here, too.' He stands there with a dumb look on his face and smiles."

"Tell you what amazes me. A lot of these new hands we're gettin' down here now are college boys."

"Yup, lot of 'em are."

"Well, if they're supposed to be college boys, how the hell come they act so goddamn dumb? I mean, no foolin', some of them are the dumbest sons of bitches I ever seen in my life."

The lowly new hand has no choice but to absorb the verbal abuse with a weak smile. Nearly everything he does is going to look stupid, which includes his constantly walking into posts, ribs, pipes, and anything else that protrudes from the darkness simply because he still automatically trusts his limited underground peripheral vision. The cap lamp light beam points straight ahead, and just because one can't see anything in the peripheral darkness does not mean that nothing is there. This is not a point easily imparted by verbal instruction, but rather part of the total underground awareness that is gained only by bouncing your head against enough posts and pipes. Nor can a new hand use much initiative to help his partner; he can't just jump on a piece of equipment and begin operating it, because the chances are excellent someone will get hurt. Working with a new hand requires step-by-step instruction all along the way, which can sorely try the patience of a miner who has enough to worry about in his heading.

"What the hell do they teach 'em in them colleges?"

"Can't be much from what I seen."

"Y'know, that can worry a man. These are the guys who are supposed to run the country and crissake all they do aroun' here is walk into posts. Gotta lead 'em aroun' by the hand."

"What I don't understand is what the hell they're up here for anyway if they went to college. The more I look at 'em, the more I see the whole country goin' to hell."

"Couple of 'em ain't too bad, but most, like you say, ain't worth the powder to blow 'em to hell."

It takes time to show a miner you're worth anything, but once favorable appraisal has been earned, the assimilation process moves a little faster. As more competence is achieved on the equipment, competence which, in an emergency, might mean the difference between the new hand's being an asset or a liability, cautious, cool inducements to the lunchroom talks are extended.

The underground lunch has somewhat more significance and importance than might be indicated by the seemingly light conversation and joking. In more conventional jobs on the surface, the lunch hour is often filled with numerous social options of where to eat and with whom, and the occasion is used to show off clothes, cars, and anything else that might create the desired image—a lot of superfluous trappings to divert attention from the person himself. Underground, there are no flashy clothes, fancy cars, and stacks of credit cards with which to impress one's peers, and all men are stripped of any ornamentation and reduced to a common level. With no outside entertainment and diversions to distract the listeners from an individual's words and actions, the half-hour lunch break can become a time of intense personal judgement and evaluation.

New hands dare not force their acceptance, but must be satisfied with a lower rung on the ladder, waiting patiently to be invited up. Only when the crew reaches unspoken agreement (usually after the arrival of another shiny new hat to fill the vacancy of crew dunce), is final acceptance shown, usually by making the new miner the object of some sort of practical joke. However humiliating, embarrassing, or discomforting, it is nevertheless an awaited event followed by raucous laughter and slaps on the back. The old timers finally address the new hand as "Pard".

My graduation was duly noted when my lunchpail was fastened to some concrete by a pair of inch-thick bolts, enough to support several tons if they had to. At the end of a shift, I had been purposely delayed from reaching the shaft on time for the ride up to the Storke. The rest of the crew, already waiting on the cage, yelled like wild men to get me to move when I came into sight. Obligingly, I raced past the concrete bulkhead, grabbed my lunchpail on the run, and wound up a second later lying on my back in the drift holding only the twisted handle. A chorus of deafening laughter echoed from the cage. "Pard, you shoulda seen yourself layin' there holdin' that handle wonderin' what the hell. . . . I ain't laughed so hard since they tied Johnny to the ladder."

Even to this day I never understood all the humor they saw in that. I guess a lot of it is in the anticipation, the knowledge beforehand of what is going to happen. Once so initiated, the new miner can and is expected to participate in the crew interaction of conversation and horseplay. The practical joking has become an institution in itself, a finely developed art, a necessary diversion to counteract the otherwise depressing mine environment. Joking falls into two categories: the simple stunts that bring a brief flurry of laughter and are quickly forgotten, and the well-planned, imaginative masterpieces that play upon an individual's personality and idiosyncrasies. One of these classics was enacted after I had been at Climax about three months.

Showing up a little early for lunch, Black and I kicked open the battered lunchroom door, surprising two old miners busily engrossed in an extracurricular project. Their initial looks of guilt turned into wide grins when they saw who we were. Spread out on the table before them was a sheet of waxpaper, two pieces of bread, and a grease gun from the mechanic shop.

"Who's it for?" my partner asked.

"Fuzzy," came the reply. "We're gonna set the Texan up with a grease sandwich."

Grease sandwiches are no rarities to mines, but although many are made, few are actually eaten. Fuzzy, so named because of his nearly bald head, was an ideal candidate for the honor. The short, chubby Texan took it upon himself to act as master of ceremonies at lunch, steering the conversation, drawing out unwilling participants, and selecting the whipping boy for the day. This often took his attention off what he was eating, even though he bragged constantly about the delicacies his good wife dutifully packed in his lunchpail. "Never the same thing two days runnin'. Never sure what it's gonna be, but it's always good," he'd say through a grinning mouthful.

For such culinary delights as grease sandwiches, trial and error had shown the best grease to be a red, opaque, petroleum-base glop that looked remarkably edible. The two chefs, still wearing all their gear and hardly able to contain their childish glee, carefully pumped the grease evenly over the bread, taking great pains not to get it too close to the edge. Just close enough to look good, but not enough to ooze out and arouse suspicion. You just don't slap a good grease sandwich together, they agreed, you got to do it right. When there was a good half inch of grease down, they garnished their offering with a piece of crisp lettuce, covered it with the top piece of bread, and stepped back to admire their creation. It looked damned good. Wrapping it carefully in the waxpaper, they placed it in Fuzzy's lunchpail.

The rest of the crew began filtering in, unbuckling belts and dropping their gear in heaps, and were informed in low murmurs of the grease sandwich that awaited Fuzzy. When most of the crew was already pouring coffee, Fuzzy made his grand entrance with his usual yell, dropped his gear, and took his accustomed place at the center of the table.

"You know how hungry I am?" he began in a loud voice. "Hungry enough to eat the north end of a south-bound skunk," he answered himself, choosing one of his favorite hyperboles from his vast repertoire. Rubbing his palms together, he licked his chops and opened his lunchpail. Mine lunchrooms are quiet, as this one was now, only when something has happened or is about to happen. Eyes shifted slyly and barely perceptible grins flashed around the table. Fuzzy immediately launched into a tirade about the poor guy from Denver who bought his old pickup and had gotten cheated in the process. "That thing wasn't worth half what he give me for it."

"How much did you say you got for it?" Good cue question.

"Why, twenty-five hundred bucks, that's what," Fuzzy proudly exclaimed, taking the cue, as he unwrapped the sandwich. From over the steaming rims of the coffee cups, all eyes were on that sandwich. He glanced down at what he was about to gnaw into, hesitated a fraction of a second, then added, "Hell, I almost felt bad taking the money," motioning with a two-handed clutch on the sandwich. Then Fuzzy leaned forward, opened his mouth to its widest gaping proportions, stuffed the corner of the sandwich in, and cleanly bit off a mammoth piece.

With his cheeks stuffed like a giant chipmunk, Fuzzy chewed heavily a few times in rapid succession, trying to get some of it down so that he might continue his speech before he lost the spotlight. It was almost silent in that lunchroom, with no sound at all except the distant, rhythmical throb of a pump in some drift and Fuzzy's chewing away. After a good hearty swallow, he continued, "But now, I told him about them valves, yessir, 'cause I'd never . . ." The words tapered off slowly and Fuzzy looked down at the sandwich, lifting the top piece of bread delicately and deliberately to peer under. A wave of color, red going to purple, flooded his face and rose to the top of his bald head as he shot a dark glance around the table. "You rotten bastards," he muttered, turning sideways on the bench making strained gagging noises trying to spit out what was left of the sandwich.

Grease, by its very nature, is not readily removed from any surface, throats being no exception, and Fuzzy was having a bad time getting the bearing lubricant out of his mouth and throat. Wracked with spasms of convulsive laughter, the kind that upsets the stomach and brings tears to the eyes, the rest of the crew had all they could do to keep from falling over. By quitting time, the story had circulated throughout the level that Fuzzy had fallen hook, line, and sinker for a grease sandwich.

By this time, I was becoming pretty well accustomed to mines and miners. I had long since stopped wandering off to become lost in the maze of drifts, and took regular turns running the heading equipment. One of the things that makes underground mining bearable is a good partner, and I felt fortunate in being paired with Dwayne Black. Perhaps it was our contrasting backgrounds, his rural Oklahoma and my urban New Jersey, that complemented each other. We also shared a mutual ability to take everything in stride. Virtually nothing went on that didn't warrant a chuckle, laugh, or sarcastic comment.

It seems a paradox that the best sarcastic humor should be born of stressful situations, military life being the outstanding example. The guys who could best adapt to the military were invariably those who could find or fabricate humor in the regimentation, the crude attempts at mass discipline, and any of the other morose and trying situations that occurred regularly. In this sense, underground mining, which is basically alien to anything a person might be used to, strikes a parallel to military life. Humor becomes the armor that can see anyone through anything, and is vitally necessary to enable one to cope with eight hours of underground darkness illuminated only by a cap lamp, a heaven of hard rock aching to come down on your head, and a battle of nerves with the cold steel mechanical monsters that wait patiently for one human mistake. That humor, however bitingly sarcastic, and the laughs it brings remain the sole tangible tie to the surface world of bright faces, blue sky, and the simple pleasures of the off-hours.

As weeks turned into months, several faces changed in the crew, with a few of the older hands leaving to work at other mines, the names of which were beginning to sound familiar. Their positions were taken by more new arrivals, mostly graduates of Wizen's "college." Personnel turnover in underground mines is extremely high, perhaps the highest in all industry, a fact that was borne out by my employee number, 11,982. At Climax, most of the turnover was because those who had never worked underground, often in a matter of days or a few weeks, decided to call it quits.

Nearly all mines operate on a twenty-four-hour basis, and the required shiftwork is a major drawback. The human body, of course, is both psychologically and physiologically geared for peak efficiency and performance during the daylight hours, but one might think that the perpetual darkness of the underground would create a sense of suspended timelessness to offset the strain of, say, graveyard shift. Unfortunately, this is not at all true, for I found myself quite aware of what the sun and stars were doing above all that rock. The normally increased fatigue experienced on graveyard shift was compounded considerably by the effects of the extreme elevation at Climax. Mountain goats may be happy at 11,300 feet above sea level, but the overwhelming weariness that besets mere humans can be awesome. The simple act of walking, lifting one foot after another, becomes a job requiring conscious effort. Months are required for the body to increase its red blood corpuscle count to compensate for the oxygen deficiency at this elevation. The resulting shortness of breath, where even the deepest inhalation fails to satisfy the demands of the lungs, coupled with the visual confines of the surrounding rock, may induce in some an acute claustrophobia. Those severely affected by this combination will last only a few shifts underground before quitting, but all, whether they admit it or not, are conscious of it to a greater or lesser degree. It surfaces as a discomfort most often felt when taking a break after some exertion, when the mind is idle. One might lie back on a piece of lagging and look up, imagining that the tightness in his chest and lungs is caused by an invisible, but distinctly tangible, force, a vice, always squeezing, constricting just enough to prevent that last bit of needed air from getting into the lungs. Then it becomes all too easy to contemplate that half mile of solid rock that stands between your lungs and that cool mountain breeze you know is rustling the needles on the timberline pines at that every moment beneath the free expanse of open sky overhead. Think about it long enough and the rock in those drifts will start closing in, imperceptibly, even in your imagination, but always closing in nearer to your chest. You look over at your partner and wonder whether he senses it too. His body, like yours, is at rest, but his eyes dart restlessly across the timber and rock in the back and tell you that his mind, like yours. . .

A dull, snapping sound comes from the back near the face. The rock is "talking."

"You hear that, Pard?"

"Yeah, I hear it." The voices, bounced back from the wet rock, have an unnatural depth, the resonance of a shower room baritone.

"Probably ought to get off our butts and get some timber up there to catch 'er 'before she all comes down."

A sigh of resignation. "I guess."

The simple act of getting to your feet is an enormous effort, but often one that is strangely welcome. You know the lifting of timbers and driving of wedges, the noise of the work, will help put the discomforting thoughts about the rock to the back of your mind. You have to keep busy. Think about that rock long enough, and the shortness of breath, and you'll stand a good chance of earning a ticket to the nuthouse.

The underground environment is too overpowering, too depressing, to allow you to think about anything outside the mine except in brief, unconnected sketches, fleeting vignettes of the other world on top. Any attempt at prolonged, creative thinking, at planning, is futile. Again, it is humor that arises as the sole substance to fill the dangerous mental void. Much of the underground humor occurs quite inadvertently, and is not recognized as such until after it has been ascertained that no one was hurt.

Black and I started every shift as usual, crowding up to the window to pick up our brass and listen solemnly to Reasoner's monotone instructions like two sinners receiving their pennance at the confessional. "Swing shift shot out the back and lost the last set," Reasoner said with his standard deadpan look. "I'll go down there with you and we'll look it over before we do anything. Don't want anybody hurt down there. Now, look over to your right."

We did. A new hand looking like a four-foot nine-inch John Denver smiled back with a ridiculous John Denver grin.

"Thought they had child labor laws in this state," Black said quietly.

"Take him with you till I figure out something to do with him," Reasoner said. "Watch him close."

We walked over to the waiting new hand, who straightened to his full height, what little there was of it. Black looked down at him and asked him his name.

"Ralph," was the answer.

"Rawf?"

"Ralph." Enunciation exaggerated.

"C'mon, Rowf."

The heading was every bit as bad as Reasoner had indicated. All the cribbing on the last set was shot out and the posts themselves leaned back at nearly a forty-five-degree angle, leaving a fifteen-foot unsupported back in front of the face. The cap had somehow managed to stay up, jammed in the timber wreckage, and resting on it was a huge slab the size of a Volkswagen. Just how many tons lay directly on that cap was a good question, but it was enough to put a visible bow in the twelve-inch-square timber. When a miner walks into something like that, he never knows precisely what it will take to bring it all down. The slightest nudge from a scaling bar may do the trick, but if the timbers and slabs are keyed into each other, powder may be necessary. Old miners have gotten old for a number of reasons, and foremost among them is the basic assumption that, until proven otherwise, the slightest nudge from anything will bring it all down.

My partner, Reasoner, and I studied the mess for a moment, talking it over in hushed tones as though the rock might hear, then inched out toward the face for a better look. Rowf, busy taking in the sights of a real working heading, craned his neck up, down, and around in wonder.

"You stay right there, Rowf," Reasoner warned the new hand. "Don't come any closer than you are."

Slight cracking sounds could be heard as the timbers protested their great burden. Step by step, keeping close to the rib, we moved forward, peering upward, every muscle tensed for a lightning retreat. Advancing past the bowing cap and beneath the unsupported back, we crept across the muck pile, looking for the key, which, if moved by a bar or by powder, would relieve the whole thing. The size of that biggest slab was awesome and was separated from the back by a gaping black crack nearly two feet wide. I could only imagine what that could do to a human body.

"You gotta put a little powder in that crack," Reasoner whispered. "Won't take more than a couple of sticks, if that, but for crissake be careful."

Rowf, standing at the base of the muckpile, had no idea what was going on. He paced around in a small circle, shining his lamp on hoses, valves, tools, and anything else of interest. Right then, of all moments, Rowf decided to break wind. The slight, muffled, popping sound went through us like a shot of electricity. Like three linemen coming off the snap, we bolted in unison away from the face in a move not of reason, but one of blind animal fear and reflex. Only Rowf, standing there flatfooted, was in our way. No one spoke when we stopped, allowing a few seconds for the adrenaline to wear off. Rowf lay sprawled in the drift where he was knocked over, then run over.

"You okay?" Reasoner asked.

"I think so." No owlish John Denver grin this time. "What happened?"

"You farted, that's what happened."

A look of great perplexity came over Rowf's round eyes. Since nothing he could see had moved, he pondered in amazement just why miners should exhibit such a profound reaction to the mere passing of gas.

At lunch the same shift someone asked what happened to the new hand, that little fella.

"Quit," Reasoner answered. "Took him to the top an hour ago."

"Didn't like it, eh?"

"I don't think he would've made it. I'll say one thing for him, though. Small as he was, when he farted, you damn sure moved."

Another game a good number of Climax miners played, one that was welcomed as a diversion, was highgrading. The origin of highgrading could be traced back to the early days of frontier gold mining when a man could easily match his daily wage by keeping his eye out for extremely highgrade ore or bits of free gold and, when he found some, pocketing them. There was no gold at Climax, of course, but a number of attractive crystals occurred in the igneous ore body that did have both monetary and collector value. If nothing else, they were unusual conversation pieces or just something to bring home to the kids. Highgrading became a competitive sport of sorts, and to the winners went the prizes of delicate, colorful, and often strikingly beautiful crystals so incongruous to the gloomy underground world of common muck.

Iron pyrite, the familiar "fool's gold," was found throughout the ore body in minute gold-like specks, but under the right conditions perfect cubic crystals would grow to more than an inch in dimension. Talc slips, seams in the rock filled with a white talc-like substance the consistancy of butter, provided excellent conditions for the development of extraordinary crystals. Should blasting happen to open a talc slip, it was a sure bet that an hour of working time would be lost while the miners dug at arm's length into the seam, scooping out the white glop and separating the pyrite crystals.

While pyrite provided a lot of amusement for all, it was the rhodochrosite, a red crystalline deposit often found lining vugs, small cavities in the rock, that drew the efforts of the serious highgraders. The rhodochrosite ranged from pale pink with excessive fracturing to ruby red, some pieces of which were suitable for jewelry and valued as a semiprecious stone. If you were so fortunate as to blast through to a vug in your own heading, the idea was to clean it out quietly and completely. When there was absolutely none left, you might make a show by casually cleaning a few choice pieces on the lunchroom table. Rumors of rhodochrosite vugs would often draw underground land rushes with miners venturing far from their headings under the pretense of "looking for tools."

Great highgraders were few and far between, but Paul Latchaw was one who had the extrasensory knack of always showing up at the right time. His job as a pipeman on a mine crew called for him to visit every heading at least once during the shift, perfect cover for a serious highgrader. He often returned to a small vug in the back of a little-used drift from which he recovered some excellent crystals, but not without risk of life or limb. Working alone, as usual, he once climbed a wooden ladder to his private vug, crawled in, and left his legs dangling. Totally engrossed in picking and chipping, he had unknowingly inched forward until he was unable to feel the ladder with his feet. Paul hung in that position for quite some time, afraid to slide out backwards, miss the ladder, and crash into the rails twelve feet below. Only the fear of missing the shift change was enough to move him to heroics. He slid out backwards, unable to see, hanging by his hands and finally dropping to a safe descent. It was all worth it, for that day he walked out with a lunchpail full of rhodochrosite, the largest piece of which was nearly flawless, a deep blood red, and about the size of a sandwich. He later sold it for $1,200.

Besides being a good miner and highgrader *par excellence*, Paul was also a practical joker of extraordinary talent, a fact I learned first hand. One of the miners in the crew had the misfortune to catch his middle finger between a drill steel and a guide. Only rarely will mining machinery give a careless operator a second chance, and this drill was no exception to the rule, smashing the knuckle and severing the errant finger. Injuries like this one and deaths are about the only thing that will dampen the collective mood of an entire crew. This is caused partially by a sincere feeling of sympathy for the injured man or, in the case of death, for his family, and partially by the cold, sobering thought that it could just as easily have happened to you.

At the end of that shift, we were riding the mantrip out, seated eight men to a metal-roofed rail car, a row of four men on each side, facing each other. On the mantrip, lamps are usually turned off or detached from the hard hat and slung over the shoulder as a courtesy to the men opposite. This causes faces to be eerily illuminated from below. There was little of the usual horseplay and joking as the outbound train rumbled through the dark drifts.

Paul, seated next to me, gave me a nudge. "Hear what happened to Jess?" he asked quietly.

"Yeah, I heard about it," I said, "too bad."

"Here, got something I want to show you. Take a look at this." Reaching inside his wet rubber jacket, he withdrew a can of Skoal. Holding it in his left hand, he removed the top with his right. I expected to see a piece of highgrade or something, but, there, nestled neatly in a cushion of toilet paper, was a finger. A middle finger. Two thoughts went through my mind; either this was the most authentic looking phony rubber finger I had ever seen, or it was Jess's finger, the real thing. Not wanting to fall for a ruse, I looked up at Paul with a blank look and gave a very noncommittal grunt. But Paul, correctly reading my mind, said with just the right touch of sadness, "Found it in their heading."

I looked back down at the finger and gave it a little poke. Damn sure felt like skin and bone. The most amazing thing about it was the dirt in the knuckle creases and under the nail. That, I told myself, is sure as hell a finger, Jess's finger. Jesus Christ that's wonderful, that's what we all need. There's enough to worry about down here without some goddamn grave robber running around collecting fingers and toes and who knows what else. I made a mental note right there not to get caught alone with Latchaw in some dark drift.

In the following days I made several discrete inquiries about any unusual quirks Latchaw might have, getting little more than puzzled stares for answers. Finally, I became involved in a serious discussion with the grease sandwich chef and mentioned the finger incident.

An incredulous grin spread over the rough, unshaven face. "Oh, c'mon, you didn't fall for that. I figured you to be a little brighter."

"Fall for what?" I asked. "That finger was the real thing."

"Well, sure it was, it was his own finger! He cuts a hole in the bottom of the can and slips his own finger in there. He don't try that crap with me 'cause he knows I'll jab the hell out of it with a powder punch. And you never seen no dead finger bleed so much in your life."

I thought about that for a while, half pissed off at being led down the primrose path, and half relieved at knowing Latchaw was not really an authentic grave robber. I also honed a hell of a point on my powder punch.

Foot by foot, day by day, the 600 Level drifts were driven through the rock, snakeline, always advancing to the roar of the drills, the concussion of the blasting, and the rumble of the muck trains. The crew worked steadily and injuries, mostly minor, were held to a minimum. Climax had a policy of immediately removing an injured man from the underground in a basket, a mummy-shaped wire mesh stretcher of which there were many stored at various locations throughout the mine. It took four men to carry a basket, and several would accompany the injured man on the motor hauling him out. With the steady wail of the siren clearing the track ahead, the miners surrounding the basket on top of the motor would stoop over as if administering the last rites in whispers, but actually were only trying to keep their heads clear of the passing caps and wires while simultaneously keeping the basket from sliding off the motor. Illuminated by the strobe-like flashes of the trolley, the motor would sway and lurch through the long straight run of the adit where Wizen's new hands would stare in awe from their manholes at what they no doubt imagined to be the passing chariot of death. It was a sight guaranteed to impress any would-be miner on his first day underground. The basket policy worked two ways; it kept phony injury claims to a minimum, but also discouraged some miners from seeking legitimate treatment because of the elaborate evacuation procedures on the "circus wagon."

Miners use two all-encompassing terms to describe most any injury. No matter what happened to a man, he had either been "slabbed" or "mashed," and only rarely would an injury, such as electrocution, fall outside those classifications. Falling rock "slabbed" you, and any other type of body breaking, crushing, or severing meant you had been "mashed." The details of just how the accident happened seemed irrelevant. Many times I asked a miner whose partner had been hauled out in the basket just what had happened.

"Mashed," was the typical answer, usually given in an emotionless voice.

"Yeah, I know, but how? What actually happened?"

"Mashed, I said," Louder, like, you got muck in your ears?

The hell with it, it was more interesting to read the accident reports anyway. All injuries, however minor, were written up in lengthy reports, the most detailed part of which theorized what the employee could or should have done to prevent the injury. Some of these were hilariously simplistic. For example: mashed finger—employee should keep finger out of equipment; slabbed shoulder—employee should bar down, check back, and not stand under falling rocks. Simple enough. Occasionally a miner, standing under supposedly good timber, would catch a slab shaken loose by a distant blast. In such cases it was difficult, even for Climax, to place blame squarely on the employee, and these accidents were labeled "acts of God." Since nothing was sacred, this was naturally corrupted into "hit by a goddamn rock."

The whole concept of underground safety was somewhat of a paradox, since it was necessary to accept a certain degree of risk in order to perform the everyday job of driving the drift. If a miner were to put absolute safety ahead of his job, there would be no work accomplished at all. Mining, I was beginning to learn, involved finding a compromise between performing your job and staying in one healthy piece. Experience and confidence, acquired over time, would help to move an individual's balance point more toward the "work" end of the scale. It was something a new hand didn't rush, as his first few months were a time of intense learning and familiarization with a new and alien environment.

One of the operations that particularly fascinated me was blasting in the headings; not so much the actual detonation of the round, but the manner in which it was checked beforehand. To determine whether a completely loaded and wired round would conduct the blasting current without undue resistance, possibly caused by a broken wire, short, or a bad cap, a blasting galvanometer was used. Referred to as a "peter meter," the unit was a simple ohm-meter with its own power source, a specially manufactured battery that was designed to be incapable of producing the minimum amperage that would detonate a cap. The round was tested simply by standing in front of the face and connecting the peter meter to the two leads, thus actually allowing electric current to pass through each of the caps, which could detonate the three hundred pounds of dynamite. If the round were good, the needle would swing all the way over to indicate minimal resistance. That was what I considered to be the ultimate trust in the quality control of the people who manufactured the caps and the battery in the peter meter. If there were a malfunction or manufacturing defect, you sure wouldn't be around to raise hell with the complaint department.

The nitroglycerin fumes from the dynamite Climax used in enormous quantities were capable of producing truly monumental "powder headaches." The fumes acted to relax the capillaries in the head, thus allowing an unusually large amount of blood to reach the brain. Skin contact with dynamite could induce the same effect. As a result, every miner carried a supply of aspirin in his lunchpail, and the company provided aspirin at the lunchrooms and other locations in the mine. Some new hands had the misfortune to learn about powder headaches from an old timer with a case-hardened sense of humor. Upon complaining of the throbbing headaches, the new hand would be told, usually by a man who wouldn't see him for the remainder of the shift, "Pard, I'll tell you a little secret about how you can stop that." Naturally the new hand would be all ears, already prematurely thankful for the helpful advice. "Now, what you do, see, is break a stick of powder in half like this, and just take a little bit of that dynamite and rub it good all over the inside of the sweat band on your hard hat. That's it, rub a little more, don't be afraid. There you go. Now you do that everyday at the start of a shift and you'll never get one of them ol' powder headaches."

That was for damned sure. The new headaches that a gullible new hand would get would start in his feet before exploding between his eyes in a blinding flash of white light and searing pain. It was a one-time mistake.

The new hand also learned other tidbits of information, subjective observations and subtle lessons from his partner's vast mental storehouse of underground knowledge.

"Pard," Black lectured in his Okie drawl as we both lay sprawled out on the timber flat taking yet another break, "one of the things you should always remember 'cause it'll help you a lot is this." He'd pause for a moment, allowing the silence to dramatize the coming words which he believed to be wisdom of nothing less than stone tablet grade. "The other shift is always a bunch of asses."

Profound, I thought. Profound, indeed. "Does that mean both shifts?"

"That's right, both shifts. But mainly the one right before us."

The ass theorem was frequently demonstrated by Black whenever Reasoner came around to check on us. The shift boss would walk into the heading and, as was his custom, eyeball the work silently before speaking. "Kinda figured you guys would've had this timbered up by this time," he'd say in his monotone, voicing his displeasure with the fruits of our labor.

Black would start in with such disappointment and sincerity in his voice it would almost bring tears to my eyes. I usually turned away to hide my grin. "Why, Kenny," Black would say, "we sure woulda had 'er done if it weren't for them asses on the other shift. Just wish you coulda seen this place when we come in here. Why it was terrible, they left shit layin' all over the place and whatever they did do was assbackwards anyway."

Reasoner would remain silent, his jaw muscles working slowly, obviously not buying the ass theorem. Then he would turn to me for confirmation of my partner's statement. I'd give him my best new hand look, dull, blank, and lifeless. Don't ask me nothin', I'm a new hand. Getting no satisfaction out of me, he'd turn to Black and say with a tone of disgust, "Let's not worry so much about the asses on the other shift, you got a lot to worry about gettin' done this shift."

Whenever Reasoner was in a decent mood, or mildly surprised by what we had done—which was not often—he would hang around and tell war stories for twenty minutes or a half hour. When he was not happy, he would spare the words and leave. On those occasions, when he had gone, Black would say with a satisfied grin, "See, if it wasn't for them asses on the other shift, I don't know what we woulda told him." I never really knew if he thought he was bullshitting Reasoner or not.

At the end of those shifts, the crew would pile off the mantrip, charge down the stairs, and fling the brass through the time windows. The shift following us would be slouched against the wall of the dry with sullen faces, waiting to brass in. Funny thing, I thought, those asses look like a mirror image of our own shift. But at that point it didn't really matter, because when we showered, changed, and headed down the hill to Leadville, the mine was the farthest thing from our minds.

CHAPTER II

LEADVILLE — MOUNTAINS, MINES, AND MOLY

Three times each day, the long line of cars signifying the end of another shift at the Climax mine winds its way down the twelve-mile-long twisting ribbon of asphalt that is Colorado Route 91 to Leadville. It seems paradoxical that the miners must drive "down" to Leadville, which, at an elevation of 10,150 feet above sea level, is the highest town of any real significance on the North American continent. It is by chance, not choice, that most of the Climax miners call Leadville home. With a population of about ten thousand, it is the only town in Lake County and the only town within reasonable driving distance of the mine. Buena Vista, a pleasant little town at the much more moderate elevation of 8,000 feet, is more than thirty-five miles south of Leadville. Although a number of Climax miners reside there, it is fifty miles from the mine. To the north, the tiny towns of Frisco and Dillon are twenty-five miles away and offer little in the way of living accommodations. To the west, over Tennessee Pass, is the New Jersey Zinc Company mine, the company town of Gilman, and little else. Residential preferences aside, the main reason for living in Leadville was perfectly obvious on the January day I hired on. Leadville simply offered the shortest driving distance to be endured during the seven months of ice, wind, and snow of the high country winter.

A few days after I was assigned to Reasoner's crew, my new partner was generous enough to offer some advice on where to stay. I remember that distinctly, because it was the first civil thing he ever said to me. It was probably quite a decision on his part as to whether to expend the benevolence and concern on a new hand. "Where you stayin' now," he asked, breaking the silence of the heading.

"Motel just south of Leadville."

"How much you payin'?"

"Thirty-five bucks a week."

"Ain't that a little high for a room?"

"Yeah," I agreed, "for just a room."

"You try the Vendome yet?"

"What's the Vendome?"

"Hotel. Down in the middle of Leadville."

The idea of a hotel didn't immediately appeal to me. The idea of the middle of Leadville appealed to me even less, although I had seen very little of it. All my time so far at the mine had been spent on day shift. This meant that by the time I got off shift, Leadville was little more than a string of neon lights smothered in a heavy gloom of deepening twilight and blowing snow.

"You ought to try the Vendome. All it'll cost you is a little time," the drawl said again at length. "Lots of hands stayin' there. Tell me it's cheap."

"How cheap."

"Ain't sure. But if there's a lot of hands there it must be cheap."

"Where is it?"

"Big red building, oldest, biggest thing in Leadville. Right on Harrison Street. Can't miss it."

After shift as I drove through Leadville, I noticed that the big, red, drab, weatherbeaten building that dominated Harrison Avenue was indeed the Vendome Hotel. It somehow lost itself among the 1900-era facades of the other buildings that fronted the main street. Viewed individually, the hotel was really a classic of gabled Victorian architecture, its stature, however, sadly diminished by the toll of the years, the faded, cracked paint, and the decades of grime.

I pushed through the double swinging doors, out of the biting wind that drove the snow in clouds down Harrison Avenue, and emerged in a large, dimly lit lobby. Walking slowly over the stone floor, past dusty display cases of old mining relics and cracked photographs of mines and miners long gone, I stopped in front of the hotel desk, a high counter of dark, heavily varnished wood backed by a maze of dusty pigeonhole room boxes and a tired grandfather clock. Behind the desk, her nearly white skin in startling contrast to the somber wood tones, waited an equally somber young lady. It seemed that in winter, the people in Leadville sport only two colors of facial skin, the pale, almost green-white of those who steadfastly avoid the bitter weather, and the rough, raw red of the snowmobilers, hunters, and skiers who find a sadistic pleasure in meeting the high country winter on its own terms. This girl was obviously not an aficionado of the great outdoors, and appeared ill-content with her lot in life in the Leadville winter.

"I'm looking for a room," I announced, my voice a hollow echo in the emptiness of the church-like lobby.

"How long?" Snow White asked, without a trace of a smile or a hint of interest.

"I don't know. Maybe a week or two. Depends."

"Climax?"

"Yeah."

Snow White turned slowly, pulling her quilt jacket tightly around her, and studied the pigeon holes patiently as though a light would suddenly appear to indicate a vacancy. When nothing happened, I asked how much the rooms were for a week.

"Fifteen. Some are twenty."

"What's the difference?"

A spark of annoyance gleamed in her previously lifeless eyes. She glared at me sullenly, as though measuring my insolence and gall. Maybe Climax miners aren't supposed to know enough to give a damn, I thought.

"The twenty-dollar rooms are bigger. And they have a private bath," she replied, as though everybody should know that.

"Well, can I see one?" I asked.

She dropped an oversize key in my hand and nodded toward the staircase. "Second floor."

The heavy bannister had been worn silk smooth and completely free of the varnish that seemed to be holding so much of the hotel together. I wondered how many years it had taken the calloused hands of the Climax miners to accomplish that feat. Little renovation had taken place over the years in the Vendome it seemed, for the once-white flower print wallpaper that covered the second floor hallway walls was now heavily yellowed and cracked. Ancient, tarnished gas fixtures jutted out of the walls at various angles, and a threadbare path had been worn in the center of the tattered red and black carpet, much like a game trail on a forest floor. Another tribute, this one to a thousand Climax boots, I thought. A door opened and a face peered out. I nodded. The face nodded back and the door closed.

I slipped the huge key into the appropriate door and opened it. The room that awaited my approval was furnished with a ragged easy chair, a somewhat newer couch, and a large desk, the recently revarnished surface of which bore the cigarette burns and other inadvertant graffiti of a half century of use. Early American, I mused to myself. The adjoining room boasted a brass four-poster, a head-high dresser, a large porcelain washbasin, and a walk-in closet. The faded wallpaper extended from the floor up to a line of molding about seven feet high, after which white paint, yellowed and peeling, took over up to the nearly twelve-foot-high ceiling, which was accented by an elaborate baroque trim of dark wood. A gas fixture in each room had been re-wired for electric lights, and another light covered by a dusty red shade dangled down from the center of the ceiling on a four-foot cord. Of all the rooms, it was the bathroom that took the prize. The bathtub was an enormous lion-footed affair with detailed scrollwork on the valves and spigot. Tubs like that weren't made anymore, and neither was the washbasin, which seemed nearly a yard across. A smaller person could bathe in the basin and swim in the tub. The toilet, not to be over-

The Vendome on Harrison Ave. Built as the Tabor Gand Hotel in 1881 by H. A. W. Tabor, it was once the finest hotel in Colorado. It still houses many miners. (Stephen M. Voynick)

shadowed by its huge sister facilities, was of such proportion that the same small person might run into serious trouble here.

I wandered around the rooms again, smiling inwardly at this living museum that was the Vendome Hotel. It was certainly not what I had expected, whatever that was, but it did have a class all its own. Wondering whether I could get used to this apartment, I recalled a bit of wisdom my father had imparted to me years ago. You can get used to anything, he had said, even hanging, if you do it long enough. With that in mind, I walked down and told Snow White I'd take it.

Within half an hour, I had moved my worldly possessions into the apartment. I opened the valves on the radiators, gaudy cast iron sculptings themselves, and after a moment of disappointing silence I was rewarded by a distant metallic banging and clattering accompanied by the unmistakable hiss of steam. The din grew into a frightening racket that gradually diminished as warmth crept into the chilly rooms. From the pair of six-foot-high, narrow, cathedral-like windows, I peered through the dingy glass into the snowy twilight outside. Below, across snowpacked Harrison Avenue, the yellow neon lights of the Golden Burro Restaurant cut through the gloom. Beyond that were block after block of weary houses, dilapidated shacks and garages, crooked picket fences, and 1950 pickups rusting in backyards beneath soft mantles of snow. Many of the structures were characterized by a pronounced absence of paint that revealed deep cracks of age in the weathered gray-brown boards. Farther to the east, almost lost in the deepening twilight, were the barren foothills of the Mosquito Range with their 13,000-foot peaks obscured in the ceiling of gray clouds. Even through the falling snowflakes, the gaunt skeletons of a hundred long-dead mine headframes stood outlined against the snowy hills. In winter, Leadville can be very dreary.

As mining towns go, the West has plenty, if circumstances of birth stand as the sole qualifying standard. Leadville, however, is one of the rare exceptions among the West's long list of legendary mining towns, nearly all of which rose meteorically to fame, riches, and glory, only to die when the mines played out and they became relegated to occasional mention in the later history books. Leadville was destined for resurrection, while her Colorado neighbors, towns like Cripple Creek, Victor, Central City, and Fairplay, became merely tourist-oriented facades of their former selves. For although time has brought tremendous economic and technological changes, Leadville still owes her continuing existence to the same thing to which she owed her birth—the mines. The story of Leadville, more than that of any other mining town, parallels and reflects the glory and the tragedy, the ups and the downs, and the tumultuous times that marked the passing of the American western mining frontier.

The highest mountain valley of the Arkansas River, where Leadville is located today, is surrounded by some of the most spectacular mountain scenery in the United States. The western wall of the valley is the Sawatch Range, the Continental Divide, dominated by Mts. Massive and Elbert, both towering over 14,400 feet. The Mosquito Range, a mere thousand feet lower, protects the valley on the east, a valley which, in Indian times, must have been a truly wild and idyllic place. It was held sacred by the Utes, who, as an utterly pragmatic people, concerned themselves only with the game, fish, and birds, and not with the bits of yellow metal that lay hidden in the gravels of the clear, cold headwater streams. The Spanish were the first to invade the valley, but, not finding the yellow metal, lingered only briefly. The first white American set foot in the high valley of the Arkansas in 1802, and for the next half century the land saw little change. It was shared by the Utes and a bare handful of American explorers and mountain men. The event that was to change forever this valley and establish a social and economic pattern that would endure to this day occurred in 1859 when a white man filled a shallow metal pan with gravel from an unknown stream. By the next spring ten thousand men had flocked to the remote valley, many making a reverse migration from the rapidly depleting gold fields of California. Although some of the better claims that first year yielded $60,000 in gold, the placers played out relatively quickly and the miners' hopes of finding another Mother Lode were not to be realized, at least not in gold. The placer activity did serve, however, to attract and hold a mining population which set the stage for further discoveries. Ironically, the miners had been plagued from the start by a heavy black sand that was extremely difficult to separate from their fine gold. This much-cursed sand, which was discarded with the tailings, was later to be identified as a carbonate loaded with silver that had washed down from nearby huge lode deposits waiting to be discovered. Had the miners been aware of this, the history and development of Colorado would have been stepped-up by fifteen years.

Large deposits of lead ore were found in 1875, which further increased mining and prospecting activity in the area. Then, in 1877, came the discovery that made the name of Leadville a household word across the country. It was silver, and the ore was incredibly rich and plentiful. Word of the monumental strike spread throughout the West and soon brought a rush of miners, merchants, and settlers, honest men willing to work for their share of the glittering wealth that was just beginning to pour from the mines.

Of the miners who arrived, there were but two types. First, there were the newcomers to the field, the optimists and the dreamers who viewed the mines as an opportunity for livelihood and as a possible road to wealth in the exciting, wide-open life of a booming silver camp. Then there were the others, the experienced miners who had cast their lot with the mines and who knew no other trade or way to earn a living, men who perhaps had already given their best years and their health to the mines, and who sadly understood that there was no fortune to be made by the underground miner, only by the mine owner.

Close on their heels came the con men, the prostitutes, and the gun slingers who would get their share one way or another, anyway they could. By 1880, Leadville had exploded into the greatest silver boom camp the American West would ever know. People continued to pour in, railroads were built, stage lines formed, and hundreds of small mining companies simultaneously competed and cheated for fortunes in silver. At its peak, Leadville was bursting at the seams with a population that approached forty thousand. It was very nearly designated the capital of the new state of Colorado.

Undoubtedly the most colorful and prominent figure of this flamboyant and often violent era was H.A.W. Tabor, who, after a humble beginning as a shopkeeper, wheeled and dealed his way to a fortune as the greatest silver baron of them all. In 1878, Tabor grubstaked two miners who promptly drove their picks into a rich vein of silver. He parlayed his profit into another mine, the Matchless, which became the foundation for his twelve-million-dollar fortune. Tabor carved for himself an immortal niche in the annals of western Americana and left his indelible mark on the city of Leadville, mainly through the construction of two buildings. The first was the Tabor Opera House, which in its heyday attracted some of the most renowned singing talent of the era to the cold, cloudy heights of Leadville. The second building was built in 1881. Like the Opera House, it was located on bustling Harrison Avenue and, at the time, was the most magnificent building in the state of Colorado. Adorned with the finest carpeting, fixtures, and furnishings, it once housed a visiting United States president, Benjamin Harrison, who pronounced it "a truly splendid building, no less luxurious than the finest of the East."

When that splendid building, the pride of booming Leadville, was built, it was known as the Tabor Grand Hotel. When Snow White checked me in eighty-nine years later, it was known as the Vendome. It is but one of the institutions, traditions, and philosophies of the mining frontier that has survived in Leadville.

The tales of early Leadville, her silver barons, and the millions of dollars that poured from her mines and smelters have all been well-documented and celebrated in books, and on stage and screen. Yet the foundation upon which this great saga was built, the mines themselves and the miners who worked them, has gone largely ignored. There is no question that the glamor of mining stops abruptly at the portal. Understandably, it is far easier to be enthralled by stacks of gleaming silver ingots awaiting shipment, glittering opera houses, grand hotels, and the diamond stickpins of the silver barons than it is by the blood and sweat of the poor miners who slaved and cursed in the subterranean darkness.

Hardrock mining in the 1880s, considering the state of technology and what was actually accomplished, was an incredible achievement. However, the achievement is little appreciated today, since only a miner who has spent his share of time underground could even begin to visualize the conditions and dangers under which these men worked. Consider first that all work was accomplished in semi-darkness with flickering candles as the sole source of light. At the start of the long ten- or twelve-hour shift, each miner was given several candles, which, if conserved, might last the duration of the shift. The candles were usually mounted in a sconce, a simple wrought-iron holder that could be driven into cracks in timber or in the rock itself. Although oil lamps provided more light, they were rarely used because the dripping problem proved a serious fire hazard in the heavily timbered drifts.

Ventilation underground was crude at best. Prior to the development of air compressors, controlled fires would often be started at the bottom of a shaft, causing the hot air to rise, thereby drawing fresh air down a second shaft. Even with the advent of the early compressors and the small lines that conveyed the air to the workings, the underground air after a blast would be so laden with dust and smoke that visability of less than a few feet would prevail for hours.

Keeping these basic conditions of light and breathing air in mind, consider then the early mining operations themselves. Drilling was a brutal operation requiring the utmost in stamina and physical coordination. The term "drilling" was actually somewhat erroneous since the holes in the rock were not made by any kind of a rotating drill as we know them today, but rather by hand sledging a simple steel. Two types of sledges were used: the eight-pound double jack, which was swung two-handed, and the four-pound single jack, which was swung with one hand. Single jacking was a job for a lone miner who held the steel with his left hand while pounding away with the right. Two and often three men made up a double jack team. One man would hold the steel, turning or "shaking" it after each strike to reposition the cutting edges, while the others pounded away with full, over the shoulder swings on the eight-pound hammers in steady, rhythmic succession. This centuries-old process was slow, tedious, and tiring, and required great concentration. One errant swing of the hammer in the dim, flickering candlelight could crush the fingers, hands, and wrists of the unlucky man holding the steel. The steels themselves were merely round or octagonal rods with a crude chisel bit fashioned on the end, and miners going on shift would carry bundles of steels of various lengths, since the bits would become dull after as little as a few inches of pounding in the hard rock. On the surface, many blacksmiths were kept occupied retempering and resharpening the piles of dulled steels after a shift. Long-handled, thin spoons were used to clean the rock dust out of the holes, a necessary step which added still more time to the already slow drilling operation. Positioning the holes was of great importance, and was determined by the most experienced man, since using the least number of holes required to break the rock would save considerable drilling time.

Dynamite, or "giant powder" as the miners referred to it, made its appearance just before 1880, replacing the black powder which had been used for centuries. Black powder was tricky to work with because the smallest flame, even a spark, would set it off, but the early dynamites were soon shown to be no less dangerous. The first dynamites lacked the chemical and physical stability of modern compounds and separated all too readily into their components of inert filler and incredibly sensitive nitroglycerin. The high freezing temperature of nitroglycerin, fifty-two degrees Fahrenheit, caused the powder to become highly sensitive to shock, and the thawing operation was the next best thing to suicide, a fact that would be readily confirmed if dead miners could talk. The new dynamites provided much more power, thereby reducing the number of holes that had to be drilled, but the miners paid for this advantage by being subjected to the fumes, both before and after blasting, which induced splitting headaches. The first blasting caps, necessary to detonate the dynamite, were manufactured of fulminate of mercury, a compound whose sensitivity rivaled that of nitroglycerin. Once loaded, the round would be timed simply by measuring the fuse length to each hole. After detonation, no less than an hour would be required to clear the air to bearable standards. Blasting, therefore, was usually done at the end of a shift to allow time for the drifts to clear.

Mucking, of course, was a back-breaking hand operation, with the loaded ore cars being pulled by mules to the shaft stations where the muck was winched to the surface in ore buckets or small cages. Mules supplied the underground haulage power for many decades, and a great number of the unfortunate animals toiled in the darkness for so long that when they were finally hauled to the surface, only for reasons of exhaustion or aging, they were found to be totally blind. Steam powered cable systems found limited use, but it was the "electric mules," introduced after the turn of the century, that finally succeeded in replacing the four-footed kind. It is interesting to note that mules were still used for underground haulage in the Bisbee, Arizona, copper mines as late as 1931.

Of all the new mining innovations, it was the introduction of the reciprocating mechanical drill that had the most profound effect upon the miners and their work. Development of these drills started in the early 1860s using steam as a source of power. These heavy, clumsy machines offered great theoretical promise for speed and economy but were finally determined to be impractical for underground use for a variety of reasons. First, of course, a steam boiler had to be constructed on the surface and the steam conveyed down into the mine through a system of insulated pipes. Flexible hoses that could withstand the heat and stress, had not yet been manufactured, and in the headings the screams from scalded miners became common. Most importantly, not even the strongest man could long endure laboring in the terrible wet heat produced by the steam exhaust in the confinement of the drifts.

Engineers, who were paid only to design the drills and not to work with them underground, then turned with considerable success to compressed air as a power source. By the mid-1880s, most mines were already using the pneumatic drills. But for each new innovation that seemed to accelerate the mining process, the miners always seemed to pay dearly, and the new drills were to be the most tragic example. Instead of breaking their backs double jacking, the miners now ruined their ears in the demonic roar of the new drills that could drill a six-foot-deep hole in hard rock in a short time. Mines could now lay off half their double jack teams to enjoy great payroll savings and lift more ore than ever before. The only apparent problem was minor: hiring more blacksmiths to sharpen the steels which were now dulled at astounding rates. The pneumatic drill revolutionized mining to the extent that much lower-grade ore bodies could now be worked economically. For the mine owners, of course, the new drills meant more ore recovered with less time and expense. They were regarded as a gift from heaven.

But if the drills were a gift from heaven for the silver barons, they were soon to become a curse from hell for the poor miners. Lubrication of the compressors and drills became the first problem to arise. The use of improper oils would occasionally cause combustion, much like a diesel engine, and a good number of miners were killed when their drills exploded in their faces. But this problem was soon shown to be miniscule compared to the strange lung ailment that began afflicting miners by the thousands. The miners who used the new drills in hard rock, which included virtually all mining in the western United States, soon developed hard, racking coughs that seemed to indicate tuberculosis, then a very common disease, especially in a place like Leadville with its extreme elevation and harsh climate. The ailment was soon labeled "miners' consumption" by the doctors of the era, a term which meant nothing about an ailment about which they understood exactly the same. Medical research, which progressed very slowly, finally pinned the cause of the lung problem on the inhalation of silica dust, the fine pulverized powder of quartz, granite, and porphyry that was produced in great quantities by the vicious hammering and rotary actions of the new and powerful pneumatic drills. The mucking and crushing of dry ore, along with the drilling, served only to stir up additional clouds of the deadly dust. Microscopically enlarged particles of the rock dust showed needle-sharp edges which scarred and destroyed the delicate lung tissues until the miner literally smothered to death. This industrial disease is now known as silicosis, and modern research has shown that the release of silicic acid was responsible for the petrification of the lung tissues. The miners themselves came up with two terms, both vastly earthier and simpler than the medical terminology, to describe the ailment which devastated their lungs. They were "rocked up" and "dusted," and more than half a century later, hard rock miners would still use those words with fear in their voices.

An interesting set of legal battles took place when several states passed regulatory laws against the "widowmaker" drills. The United States Supreme Court, reflecting the attitudes of the 1890s, finally ruled that if a man accepts a job in a hazardous industry, it is simply a matter of tough luck if the hazard catches up with him. The reasoning was that the companies had not forced the man to take the job and, therefore, both their legal and moral responsibilities were fulfilled when the weekly payroll was handed out. Furthermore, the Court went on to say that the company was in no way obligated to alleviate the hazard if doing so would cost money. All this legal action served only to assure that the rich would get richer and the miners would continue to get dusted and rocked up and cough themselves to death in the white cloud of rock dust that surrounded the powerful new drills.

At some point back in the frontier mining era, a basic philosophy was formulated among the miners. It has persisted until this day. Why, one might ask, since the miners were quite obviously aware of the silicosis problem as well as the omnipresent threat of injury and death, were there always men ready and willing to work in the mines? Simple financial need, the responsibility of supporting a family, certainly played a great part. But beyond that, there was apparently the recognition of the physical challenge of the job coupled with the almost childlike satisfaction and fulfillment that comes with taking the chance and trying to beat a difficult game on its own terms. Considered in that light, men have come to think of mining as something almost akin to a most serious sport, a game of sorts, and it is this attraction, conscious or not, that is referred to when people speak of mining as "something that gets in your blood." The inordinate dangers of mining then become accepted as an inherent part of the job. The miners never questioned whether greater remuneration might be warranted. Having fathered this self-image, miners have always unknowingly or uncaringly worked to strengthen it, and mine owners and management have always been quick to take advantage of it, a labor relations quirk that holds true today.

About 1890, as the Leadville silver mines continued to pour out their gleaming treasures, and as the silver barons got yet richer and the miners continued to get rocked up, a system was devised to divert some of the compressed air down a longitudinal hole in the drill steel, thus automatically blowing the holes clean of rock dust. In that respect it was a success, but it created such dense clouds of rock dust that the miners positively refused to use them. One of the major breakthroughs of all time in mining technology was accomplished when, instead of air, a stream of water was sent down the steels into the drill holes. The water served to cool the steels, flush the holes, and form a mud which acted as an effective wet grinding compound to further speed up the drilling process. Most importantly, the water eliminated the deadly clouds of rock dust which had been destroying the miners' lungs. Anti-dry-drilling laws were finally enacted in all the states. The mining companies wholeheartedly supported this legislation, not for the sake of the miners' health, but for the simple economic reason of the greater efficiency of wet drilling.

Although several different European ethnic groups worked on the western mining frontier, it was the Cornishmen who had by far the greatest effect on early American mining technology and tradition. Greatly in demand because of their long experience in the tin mines of Great Britain, they were first drawn to the United States by the work available in the Illinois lead mines. From there, it was a short step to the booming gold and silver mines of the western frontier, where they picked up the name of "Cousin Jacks." The true origin of the term has been lost in history, but one story has it that since John was a common first name for the Cornish, whenever a mine superintendant decided to hire more miners, every man on the crew would volunteer his cousin

Jack. The Cornishmen brought with them a pride in their work which has survived to this day. Their general pattern of drilling, blasting, and timbering, which they had developed almost to the level of an art, produced a uniformity and conventionality of technique in the western frontier mines that is still easily recognizable in modern drift development.

The Cornishmen brought with them two institutions which are very much a part of underground hardrock mining today—contract mining and highgrading. Contract mining, in its simplest terms, was an arrangement between the miners and the company. Under the arrangement, pay was not determined solely on the basis of time put in, but rather on what was done. At first, teams of miners would submit bids on certain underground projects like drift development or ore extraction. If the bids were accepted by the companies, the miners went to work with their profit determined by how well they could cut their own cost and time. The companies usually issued such contracts only when they had reason to believe that the planned project would be unusually difficult or dangerous. They hoped, of course, that the miners would absorb the loss should one occur. The miners relied heavily on their ingenuity and expertise, using every technical and economic shortcut possible, even to placing themselves under greatly increased personal risk or ruining the underground workings for those who would follow them. They would blast with insufficient powder, install the least bit of timbering they thought they could get away with, and, on occasion, cut sections from the superintendent's measuring tape. Since a contract meant money, the miners knowingly assumed incredible risks, doing little that was not absolutely necessary, and the increased rates of death and injury readily verified it. More than just a few miners walked into their headings at the start of a shift to find the rib dusted heavily with quicklime, a silent memorial to his opposite on the other shift whose luck had suddenly run out. Contract mining is still a viable system today, and its survival may be attributed to the mutual economic need and the attractiveness of the battle of wits.

Highgrading reached its peak in the gold mines of the West, although it was practiced with considerable success in the silver mines also. The miners believed it to be a natural and traditional privilege that helped compensate for the dangers of underground mining. They reasoned that if a slab could break their heads, any highgrade which happened to come down with that same slab should without a question go into their pockets. Much stronger than any moral justification was the financial motivation, since, prior to 1900, miners earned only three or four dollars for a ten-hour shift. An ounce of gold the size of a marble was worth sixteen dollars, nearly a week's pay, and in the darkness and danger of the underground only a fool would place himself above highgrading. The Cornishmen, who received the credit or blame for most of what went on underground, devised ingenious methods of sneaking their highgrade out of the mines under the alert eyes of the guards and superintendents. Hidden pockets, compartmented lunch-pails, and every bodily oriface were potential hiding places in the battle to outwit the company. The mine, companies considered the practice outright theft, of course, since many profit margins were made or broken by the gold that was spirited out. Although many miners stood trial for highgrading, few were convicted, thanks to the juries which were packed with their partners who knew damn well they might be in the same dock tomorrow.

The conditions under which the miners worked certainly fostered no company loyalty. It is easy to understand why, in the booming frontier days when there might be dozens of good working mines in a single district, nearly all the miners were tramps. It was jokingly said that a mine required eight men to handle a two-man job underground, since there were "two a-coming, two a-going, two a-rustling, and two a-working." The vital conversations at the lunch break dealt solely with "who's a-hiring and when," and a company never really knew whether a man still worked for them until he actually showed up for a shift. Entire mining districts would be devoid of any good working miners within days of an unfounded rumor of a bigger and better strike over the hill.

Such was the picture of the western hardrock miners and the mines in which they labored in Leadville and all over the West in the late 1880s. Leadville was now considered a metropolis of the Rockies, and the silver still was shipped out in gleaming ingots. But like all western mining boomtowns, disaster lay just around the corner, and, for Leadville, it came in 1893 with the repeal of the Sherman Silver Purchase Act. This act had required the federal government to pur-chase annually at least 4.5 million ounces of silver, nearly the total national production at the time. In the ensuing panic, 580 banks and sixteen thousand businesses across the country failed, and countless mines, mills, and smelters all over the Rockies were shut down, most never to open again. Leadville, a city that was literally and figura-tively built on silver, had its foundation quickly and cruelly ripped from beneath its feet. Thousands of miners, instantly without work, left the dying mountain city to tramp to the few remaining productive gold camps in the West. As fast as Leadville had been born, it had died, and when the twentieth century had arrived, Leadville was a mere shadow of its former self.

Back in 1879, when the mere mention of Leadville would excite people a thousand miles away, when the magic word was silver, and when everything Tabor touched seemed to turn into a fortune, a lone prospector named Charles Senter, who by nature shied away from the hordes of wild-eyed silver seekers in Leadville, was busy tapping his rockpick into an outcropping far above timberline. The rock which interested him was white and shot through with dark streaks which smeared and felt greasy to the touch. The best assayers and mineral men of the region were puzzled by the samples. The collective opinion was that Senter had discovered a form of graphite. One can only imagine what went through Senter's mind as he pondered the fact that only eight miles away everyone who stuck a pick into the ground was striking silver, while his lot was graphite. Nevertheless, the prospector went through the ritual of staking out his graphite claims on a mountain named Bartlett—a mountain destined to touch the lives of many thousands of people who, years after Senter's death, would work not on it, but under it.

A quarter century after Senter's discovery, enlightened scientific minds now pronounced the white rock with the black streaks to be molybdenite ore, the primary mineral of an element that was in no way connected with any form of graphite. The trouble was that molybdenum, recognized as an element only in 1833, was still just a laboratory oddity, meaning Senter had a mountain full of something nobody knew what to do with. Perhaps he was able to find consolation in at least knowing what his strike really was. It was the Germans who developed the first practical uses of "moly," combining it with steels to form alloys especially well-suited to the severe stresses encountered in armament components. That fact, coupled with the sound of the guns in World War I, soon turned American attention to Bartlett Mountain. The Climax Molybdenum Company was formed and commercial mining began at Fremont Pass in 1917. "Electric mules" now pulled the ore cars that were filled by Leadville miners, who by this time wore carbide lamps on their cloth hats. A short shutdown followed immediately after World War I, but in 1923 the rumble of muck trains could once again he heard in the maze of drifts beneath Bartlett Mountain, a rumble that hasn't stopped since.

Today, Leadville is just as dependent on the miners' paychecks for her existence as she ever was. The difference is that, instead of working a hundred small mines, the miners draw their pay at one enormous mine, Climax, and two smaller ones. Leadville's economic and social dependence upon Climax cannot be overestimated, as the influence of the huge mine touches every facet of life in the town. The peak hours of street activity and traffic, for example, occur during the

Homes in Leadville, Colorado reflect their Victorian heritage. When they were built Leadville still relied on silver as its life-blood. (Stephen M. Voynick)

The Pioneer Club in Leadville has been a favorite of miners for drinking and other entertainment since the 1880s. Mt. Massive stands in the background. At 14,418 feet, it is the fourth highest peak in the Continental U.S. (Stephen M. Voynick)

shift changes, three times a day, every day. The time of day itself is more often expressed in relation to a particular mine shift—day, swing, or grave—than it is in the conventional system of hours. It wasn't long before I found myself saying things like, "I'll see him when he comes off days," or being told that, "after grave is a good time to hit such and such a bar."

About an hour before the start of each shift, the shuffling of feet would start in the halls of the Vendome and, like clockwork, there would be a knock on the door. I'd open it and look into the sleepy face of Pat Kelly, another of Reasoner's miners who also lived in the hotel. Pat could be very talkative before swing when he was reasonably awake, but before day or grave he could only stand there with his "let's get it the hell over with" look. Each shift, of course, would begin with the ride up the hill to Climax. Although we alternated cars, I much preferred mine, which had a working heater and a radio, and also a better chance of starting if it weren't too far below zero. Falling in line with the traffic on Colorado 91, we would listen to the local radio station, aptly named KBRR, and get an earful of local news and country and western. Before grave we could get the distant nighttime stations with the news of the world, which seemed so irrelevant to the world of Climax and Leadville.

I grew to hate the idea of going underground on days, particularly when it was a clear morning and the first light from a low rising sun would color the distant snowy peaks of the Sawatch Range a delicate, warm red fifteen miles away in the crystal clear air. The clarity of the high country air above 10,000 feet would almost be startling and each pine, rockslide, and snowdrift on the surrounding slopes could be seen with such sharpness and detail as to produce an illusion of fantasy. There were many days in the winter, however, when we were glad to reach the mine dry to get out of the rotten weather. Why there weren't more accidents going up and down that damned hill was a mystery to everyone.

More often than not we would drive in silence, each of us lost in his own thoughts. Occasionally Pat would bestow upon me some of his personal philosophies of life. "You know," he'd begin, sliding back in the seat, a wry grin spreading across his face. It was that grin which foretold of the imminent disclosure of a comment or question that might probe the very meaning of life itself. "You know, more and more, I don't think it's worth it."

"What isn't worth it?" I asked.

"All this crap," he answered emphatically, with a sweeping gesture at the nearing brown-orange scar in the mountain ahead that was the Climax glory hole. "All this breaking your ass and running around in holes in the ground getting hit in the head with rocks and run over by trains."

Pat was somewhat prone to exaggeration, because he had never been hit in the head with a rock or run over by a train, not yet, anyway. And he certainly didn't have a reputation for breaking his ass under any circumstances.

"I've been here a year and a half now," he went on. "When I got here I told myself it was for six months, time enough to make some money to go back to school. More than a year now, and I'm still here. That's what worries me. I don't think I know why I'm here."

I only smiled, not feeling up to the mental exertion of debating the degree of self-control we had over our destinies.

"Alright," Pat said, not about to be put off so easily. "Why are you here? You came out of New Jersey or New York or one of those goddamn places out there and now you're up here in the middle of nowhere." His voice rose as he went on, "Go on, tell me. What the hell are you doing up here?"

Here it comes, I thought. Positively, with slow, measured words, I stated confidently, "I'm here to get some money and get the hell out." Of course I know why I'm up here.

"Sure, sure. That's exactly what I told myself when I started. Exactly. Well, I got the money and I'm still here. And you know what else? You're no more of a goddamn miner than I am. You don't like this running around in the dark dodging rocks, for crissake. The whole thing is, you don't know why you're here either, whether you want to admit it or not. And I'll tell you one other thing. You're going to be working in this place long after you think, just like me.

"You're not going to get out of here until you figure out why you're here in the first place. This is just a time-out up here. Halftime. Where the world stops. But it's funny, because you just don't get back on the world again that easy. Oh no, you got to make a mental effort and fight to get out of here because something's got you." A pause. Then, with an air of finality, slowly, and with precise enunciation, Pat thoughtfully condensed the lesson of the morning into a single sentence. "When you really understand why you came here in the first place, then, and only then, do you know it's time to leave."

I didn't know whether that speech was for his sake or mine, or whether it was the heater or the gnawing idea that Pat might be right, but I needed the rush of frigid air that hit me in the face when I opened the car door in the parking lot. Walking through the powdery snow to the dry, Pat told me with an amused smile that he would discuss the whole matter with Reasoner when he got to his heading. I could just imagine that.

"He thinks I'm crazy, anyway," he mumbled.

At the end of that shift when I brassed out, I learned that a slab of rock had imparted its own philosophy on Pat. It was nothing serious. The precautionary X rays taken after he had been hauled out showed only bruises. I wondered whether he was any closer to understanding why he was here, since falling rock has a not-so-subtle way of providing extraordinary insight into the study of such questions.

The circumstances that had brought Pat Kelly to Leadville were interesting, but not unusual. After attending Colorado State for a year or two, he had encountered some problems with women and money. The way he explained it, it was the flight from his wife, more than the need for money, that drove him into the mountains. For his first six months at Climax, he lived in an Army surplus tent at 10,700 feet on the Ten Mile drainage north of the mine, and maintained only a post office box in Leadville for contact with the outside world. He was obsessed with his low-profile, secretive life style, noting with confidence that the bloodhounds would probably die out at 9,000 feet. It was only the extended sub-zero weather of February and March that finally forced Pat into the steam-heated luxury of the Vendome.

Nearly everyone working underground at Climax was a character indeed. The sane and the stable, the pillars of society at normal elevations, were certainly not the ones who answered the ads to work underground in a mine on top of the Rockies. The ones who did make it to Leadville and Climax were the lost, the self-styled adventurers, the wanted, the unsettled, and the curious, each in search of his own Holy Grail. It was, in fact, a situation that closely paralleled that of the silver boom in Leadville a century ago. Even then, it was not those who were content in their station and their security who slogged up a frozen mountain in search of both the literal and figurative vein of silver. In many instances, it was those who were still trying to determine what it was they should be searching for. Seek and ye shall find. But what shall I seek?

The new hands that beat a steady stream up to Leadville in 1970 represented a cross-section of young America, one that mirrored the scars and unrest that had been created by the war in Vietnam, and one that questioned and doubted many of the established values of society. Although the Climax employment ads emphasized the "no experience necessary" aspect, it was not just the unskilled or the uneducated who answered them.

Although I had only been at Climax for several months, I had picked up the mannerisms and expressions of the miners, wore my belt and gear in that certain way that connotes experience, picked up a good layer of crud on my hard hat and lamp, and repeatedly washed my work clothes into that nondescript brown-gray that allowed one to blend into the bustling floor of the dry at shift change without so much as a second glance. No way was I any kind of an experienced miner yet, but I sure as hell looked like one. On Mondays, before the start of the shift, I would watch in amusement as the new hands, fresh from their week of orientation and safety training, would appear at the top of the stairs leading down to the dry floor. One must be around a while to fully appreciate just how virginal a green hand can look on his first day on the job. Clean clothes, new boots, a shiny hard hat, and a line of questions at least as dumb as the ones I had asked a few months earlier.

Reasoner's crew picked up about one new hand a week, an event that always brought a look of mild anguish to his stone face. Our new hands included an ex-cop from western Colorado who had yet to realize he was no longer fighting in a Marine recon outfit in Vietnam, and an ex-mining engineer whose last job had been in Australia. Both of them were politically ultra-conservative and actually qualified as right-wing extremists. Both of their rooms in the Vendome could have passed for an insurgent's field headquarters, with the variety of .45s, .38s, rifles, shotguns, and knives. They even possessed such specialties as brass knuckles and a harmless-looking wooden rod four inches long and an inch thick.

"Looks harmless, doesn't it?" asked the engineer, with a knowing look in his cold eyes as he tenderly cradled the worn piece of wood. "The Philippine guerillas used these all the time. See, you can use it to reinforce your fist, or," pausing to raise the instrument in front of my face so I wouldn't miss a thing, "use it to cave in somebody's head with the ends that stick out past the sides of my fist. The beauty of it is that you can carry it on the street and no cop can say a damn thing."

"Beautiful," I agreed, checking my watch as though I had to go somewhere. Anywhere.

The ex-recon-cop, however, made it plain that, rather than turn to gimmicks like that, he would rely on his chrome-plated service .45 "when the time came." Both maintained complete libraries of volumes dealing with the killing arts, either of which, they said, I was free to use.

Two others came in about a week apart. Dan Larkin and Steve Campbell wound up as partners, eventually in the same heading. Steve had been a helicopter mechanic in Vietnam, and, upon getting his discharge, learned that the market for that particular skill was thoroughly glutted. Dan, with a B.A. in English from an eastern college, apparently wasn't prepared to pursue the professional options open to such an education, options requiring him to "rub shoulders with the bourgeoisie." There was no limit to the backgrounds and educations of the Climax new hands.

College degrees, or at least college attendance, were surprisingly common for many of the new men who came to break rock in the dark at 11,300 feet. Of course, there were also those who lacked even a basic high school education. In the underground, there was little if any difference in the work performance or learning capacity of these men. The amount of time required to turn out a miner who could drive a heading on his own without undue supervision hinged mainly on a man's willingness and readiness to accept a measured degree of risk to perform a job. That level would vary with each individual, and had to be determined by trial and error before any degree of confidence could be realized. As expected, most of the injuries occurred to new hands before they had determined their own particular risk levels. A miner who was totally defensive and concerned only with avoiding personal injury and keeping out of harm's way in general would never accomplish a bit of mining. Conversely, a hero, hell bent for recognition, a "super miner" as they are called, is almost certainly headed for early retirement. The unspoken rule of underground mining, agreed upon by both the companies and the miners themselves, requires achieving a compromise between the two extreme positions—getting the job done and keeping yourself in one piece. If you are not willing to accept this basic premise, don't bother trying the job.

I recall one shift, when that whole concept of accepting risk and drawing the line was presented quite clearly. When Black and I brassed in, Reasoner told us that the previous shift had shot at quitting time and that they "weren't sure about the back." "So watch it," he said in a level voice, and with a serious look in his eyes that emphasized the point far better than words could have.

We had been enjoying pretty good rock in the heading, but when we looked at it that day we saw the situation had suddenly changed. The last ten-foot round that had been shot out revealed some of the rattiest ground I had seen in my brief time underground. A lot of the lunchroom stories were about bad rock, but talking about it over coffee and getting ready to work under it are very different. At the face lay an enormous muck pile, more broken rock than I had ever seen from a single shot. After we dropped our lunchpails and bits in the rib and approached the last timber set, we could see where all that rock had come from. Stepping cautiously from beneath the cap, I detached my lamp and aimed it straight up. There was nothing there but a gaping, black cavity where the back had been unable to support itself. At least twenty tons of rock had come crashing out to make that hole. Neither of us said anything, and I had that disconcerting gut feeling that this was going to be a long eight hours. As if to confirm that idea, a deep, sickening groan emanated from the cavity in the back. Not loud at all, it still possessed all the ominous qualitites of a deep, low growl from the throat of a large predator. Suddenly there was a gray flash, a barely visible downward movement, followed by a series of heavy thuds as another ton of rock let go. The two of us walked back off the muck pile and over to our lunchpails, where we sat down along the rib and poured some coffee.

"What do you think, Pard?" I asked at length.

"I think we shoulda stayed home."

I thought so too, but it was a little late for that. We sipped the coffee slowly, very slowly, then poured another cup. Every once in a while we could hear the rock talking, reminding us that we could only drink coffee so long. Sooner or later, we were going to have to do something about that rock. As far as I was concerned, they could pour that son of a bitch full of concrete and start driving off in another direction through better rock. But things didn't work that way and I knew it. The light that appeared in the darkness was Reasoner. He scrambled up on the muck pile and studied the back, then slid down and lit a cigarette.

"Gotta be rockbolted," he said flatly.

Just thinking about pounding that rotten rock with a jackleg was enough to make the coffee taste like drain cleaner. Black apparently felt the same way, because there was no move to set up a drill.

"Ought to get it over with before she gets any worse," Reasoner said. He left unspoken the fact that it was going to be bolted.

It'll take half the shift to bolt that back, I thought, looking up into those slabs the whole time. I visualized the weekly paycheck on one side of an imaginary scale, and a slab with my name on it on the other. It took very little imagination to picture what a human body would look like if it caught one of those babies. I thought of an old timer I had seen on the surface whose arm, between his shoulder and his wrist, looked as if it had four elbows in it. I was satisfied with the factory issue of one per arm. 43,000 tons of rock get hauled out of this hole every day, I contemplated, and if Number 11982 gets smeared, they're still going to haul out 43,000 tons of rock, along with one hundred and eighty pounds of fresh meat. The justification of the risk seemed to be getting smaller the more thought I gave it, but the possible consequences loomed ever larger. Another ton of rock slammed into the muck pile. Before the dust had settled from that one, Number 11982 had made his decision. But he never had to tell anyone about it.

Reasoner broke the heavy silence first. "Okay, you two set up the drill. I'll drill it." No doubt about it, he was perceptive indeed. It was simply an example of the variance of how much risk we were each willing to assume. There was no question of company loyalty, of which there is damned little among miners anyway, or of trying to maintain production and scheduling, or of any other rationalization. It was simply where you drew your own line.

Reasoner dragged the drill up on the muck pile and positioned himself in a small pocket near the rib that would offer at least some protection. When he was ready, he gave the drill a few short bursts to see what the initial vibration would bring down. Small stuff. In a couple of hours he had the back drilled and bolted. Some good sized stuff came down while he was working, but Reasoner had the expert miner's knack of being in the right place at the right time.

When he was through, all he said was "Okay, now get 'er timbered up." He walked down the drift, the glow of his lamp disappearing slowly in the darkness. Black and I just looked at each other. When we finished the mucking, we brought up the timber flat. Even knowing the rock bolts were in did little for my confidence, and in the final hours of the shift the timbering itself got hairy, with good-sized chunks of head breaker coming out of nowhere. The 43,000 tons of rock were hauled out that day, too, just as it always was. At the end of shifts like that, where you spent hours just waiting for something to happen, the first light that materialized out of the blowing snow on the drive toward Leadville would radiate more than its usual warmth and welcome. That light was red neon and it spelled out COORS.

The bitter winter weather that had dominated life in and around Leadville when I hired on had continued unrelentingly through February, March, and well into April. Even in May, snow every day was more the rule than the exception, and the temperature in the mornings still dropped to near zero. I started to wonder seriously whether Leadville really had a spring and a summer as I had heard. In May, both Denver on the eastern slope and Grand Junction on the western slope would be fully immersed in the green and warmth of spring. On trips to Denver in May, there would be the wonderful sensation of freedom to be able to walk about in a shirt, but the return trips were depressing. Spring would fade away with the increase in elevation and, approaching Leadville, everything would still be gray-brown and drab through the incessant snow showers.

It was the last week of May before any appreciable greening could be detected in the brush that covered the flats of the Arkansas River. Stream levels began rising as the snow cover melted away in the days which sometimes got up to fifty degrees. About the first of June, the aspen groves began greening, bringing a hint of life to the somber dark green of the pine forests that blanketed the slopes of the mountains. On the floor of the forests and meadows, tiny flowers appeared to lend bits of brilliant color to the lifeless earth that emerged from beneath the disappearing snow. Just when I was certain there would be a spring as promised, three final days of bad weather dumped nearly eight more inches of snow in early June. When Climax stopped tabulating the official snowfall at Fremont Pass on June 1, it had recorded 262 inches since September 1 of the previous year.

No one, unless he or she has lived through a Leadville winter, or one similar in some other garden spot, can begin to appreciate what spring can really mean. Too often, at lower elevations, the arrival of spring is lost in the gradual changes that occur unnoticed and, therefore, is marked and remembered only as a calendar date. Although spring comes late to Leadville, its arrival is accompanied by changes so profound, so sudden, and of such striking visual grandeur that one can only ponder the infinite forces capable of reversing the course of the long, severe high country winter. The end of June in Leadville is as beautiful a time as there is anywhere in the world. Both the forest meadows and the higher alpine meadows above timberline are thick green carpets that explode in magnificent displays of wildflowers that stand out brilliantly against the snow-covered peaks. The aspen groves, with their startlingly white, straight trucks, rise up from the flowered forest floors to umbrella into ceilings of green-yellow leaves, each leaf gently quaking in the warm breeze. If anything, the alpine summer made it harder than ever to drive up that hill to Climax, knowing you were going to spend eight hours in those dark drifts of cold rock where there is no spring, no changing seasons, just a perpetual damp darkness.

During the long winter, much of the surrounding area is inaccessible to anything but snowmobiles or snowshoes, since it is an utter impossibility to keep the secondary roads open during the seven months of drifting snow. But as the roads open, so does life in a sense, since one is no longer bound to the plowed roads and the major highways. A world of rushing trout streams, forest and mountain trails, and the little-visited places that offer both great beauty and complete solitude suddenly appear to alleviate the dreariness of a Leadville winter. One of those areas that opens with spring is as much a part of Leadville as the Vendome and Climax. Many people, however, view it in a different way, depending, I guess, on how one views life itself. If one is utterly practical, the nearly twenty square miles in the foothills of the Mosquito Range immediately east of Leadville are nothing more than a junkyard of mammoth proportions. But those with a freer imagination, and a yen for the romantic perhaps, would feel that the ghosts of Leadville's past still haunt the vast ruins, and that the wooden headframes,

Ruins of the million-dollar silver mines still occupy
a large area of the foothills of the Mosquito Range,
located just east of Leadville. (Stephen M. Voynick)

An old Cornish headframe still stands atop a hill near Leadville, a monument to the miners who made it all possible. (Stephen M. Voynick)

weathered hoist shacks, and decaying tramways stand as eerie monuments to the men who worked them a century ago. Abandoned mines with enchanting names like Silent Friend, Fanny Rawlings, Maid O'Erin, Little Ellen, and a hundred others, names that somehow bring a personal and human quality to the cold, lifeless ruins, still seem to offer a vague promise of riches. They rest amid frayed cables, battered ore buckets, and rusted ore cars on the slopes of the scarred, tunnel-ridden hills. A graveyard silence prevails here as the mountain wind whispers through the bleached timbers an eloquent soliloquy to the old miners, who, it seems, were suddenly and inexplicably called away from their work, leaving their tools and dreams for someone else who never came.

There is a limit to romanticizing anything, of course, and a miner of today can look at the ancient equipment, the rickety headframes, and the broken shovels, and can visualize rather well just what those poor bastards went through in those holes. They suffered the long shifts by candlelight, the bad air, the crude equipment, the tuberculosis, the silicosis, and the miners' constant companion, the falling rock. All of it put together composed a grand scheme enabling the rich to get richer and the history books to fill in later years with glorious accounts of the courage and daring it took to establish and develop the western mining frontier. The Leadville ruins are indeed a monument, truer and more meaningful than any concrete statue or brass plaque could ever be.

The technological contrast between the state of the art of the mining frontier and the modern operation at Climax today is paralleled by an architectural contrast that may be seen on Harrison Avenue. The buildings that front Leadville's main street are an exhibit of architectural progress that ranges from the 1880s to the present. It is a clash between the past and the present with the past persevering, for few truly modern buildings have challenged the Victorian styling of the boom days. The austere Tabor Opera House, vacant front windows staring out like empty eyes, begins a procession of pre-1900 hotels and commercial buildings that ends four blocks later with the decaying splendor of the Vendome, once the Tabor Grand Hotel, the pride of Colorado.

Most of the houses on the streets immediately off Harrison Avenue are a half century old and styled with the steep roof peaks, gables, and elaborate latticework that were popular decades ago. Many tourists assume the peeling paint and weathered shingles to be the manifestations of abject poverty, which, in reality, is hardly the case at all. It is just that Leadville residents are utterly practical in their philosophies, and cater little to vain pursuits. If a fence is standing, a coat of paint won't make it stand one damned bit straighter. There is a small group of newer houses on the western limits of town that lends a touch of normal suburbia with their fresh paint and trimmed lawns. These homes, all remarkably similar, are clustered closely together as if for protection against whatever it was that afflicted the rest of Leadville. The rock-steady employment levels at Climax do not serve to encourage any significant new home construction, and, instead, have fostered a sustaining condition.

Sustentation has always been a key word in Leadville, apparent both in the functional preservation of its frontier architecture and in the conservatism of its residents. Located off the main east-west highways that breach the Rockies, the town has always contented itself with a state of semi-isolation and an existence at the fringe of the mainstream of contemporary American life. Change comes slowly here, and has always been met with resistance. In fact, the new hands at Climax more often end up adopting the Leadville life style and philosophy than contributing any real change or social innovation to the town.

But such is not always the case. On occasion, a lasting social phenomenon would take place. Just such an event occurred in 1970 when a new hand showed up in the dry on a Monday morning. Aside from the fact that he was a typically dumb new arrival, this one wore shoulder-length blonde hair, and his shiny hard hat was adorned with a red, white, and blue American flag pattern. The usual din in the dry dropped abruptly to a low murmur as this guy made his way to a window to sign in with one of the other mine crews. Reasoner, I'm sure, his stone face framed in the time window, breathed an audible sigh of relief when he discovered this new hand had nothing to do with his crew. I never learned his name, since most everyone was soon referring to him as "Captain America" after the *Easy Rider* motion picture character. Initially, many of the older hands referred to him as "that fruit hippie sonuvabitch" with a vehemence in their voices that showed they meant it. But to the surprise of many, Captain America was soon carrying his share of the drilling, blasting, and mucking like everyone else. His first weeks at Climax weren't easy, and a less determined person would have packed it in. Acceptance was reluctant on the parts of many miners and came slowly, but Captain, as he was called for short, was the first of many like him who came in the following years.

Another notable social contrast concerns the town of Aspen, which is only about twenty-five miles in a straight line from Leadville over the Sawatch Range and the Continental Divide. Highway distances are considerably more, considering the winding, indirect mountain roads. The most direct route is a spectacular drive which crosses Independence Pass at 12,100 feet, the highest paved pass in the country. Both Leadville and Aspen were established as mining camps, and both rose to fame and fortune on the crest of the silver boom. During the boom era, the two towns maintained close ties, as their economies and industries were the same. The only real difference was Leadville's considerably greater size.

The silver-based economies were both suddenly shattered forever with repeal of the Sherman Silver Purchase Act, which withdrew federal support from the entire silver mining industry in the West. Both Leadville and Aspen became near ghost towns as the citizens filed out to relocate in more viable communities. Many of the miners headed for the Cripple Creek district of Colorado, where one of the greatest hardrock gold fields ever discovered was just opening. After the turn of the century, Leadville dabbled in mining, continuing its tradition, but Aspen, 2,000 feet lower in elevation and enjoying a correspondingly milder climate, turned to cattle as the basis of the economy for the people who remained. It was here that the paths of these two towns, once so similar, began to diverge. With the opening of Climax, mines and miners returned to Leadville and everything in the two-mile-high city was pretty much as before. Decades later, while Leadville was content with its mines and miners, it was decided that strapping boards on one's feet and sliding down the snowy slopes at Aspen could be developed into an industry. By the late 1950s, it became apparent just how right they were. I imagine there was a frantic cattle drive to get the cows the hell out of town to make room for the chalets, ski slopes, restaurants, and resorts that mark the town of Aspen today.

The result, of course, was a steady stream of new people, generally more educated and sophisticated, who came to ski at Aspen. Their particular culture and society took roots in the Rocky Mountain soil, and Aspen is today a center, real or imagined, of the arts and of intellectual enlightenment. However, it is the belief of many that the residents of that fair town, in their zeal to further that image from a commercial standpoint, have erected a number of notably phony facades. It was only natural that the counterculture should also thrive in Aspen, placing the once near-twin cities of Aspen and Leadville at opposite ends of the social and cultural scale.

Leadville continues to adhere to the time-honored conservative principles and work ethics of the frontier days, while Aspen revels in its liberalisms and avant-garde styles of speech, dress, and thought. Although there is no open antagonism between the two, neither is there any love lost. The mere mention of Aspen to a Leadville old timer who has put in twenty or thirty years in the hole will probably stir mumblings about "those hippie fruits over there," and conjure up a mental image of Captain America leading a band of followers over Independence Pass to take over Leadville. Conversely, the mention of Leadville and her miners to an enlightened Aspenite brings a down-turning at the corners of the mouth and a look of either disgust or pity.

The two towns are further polarized by an annual event in Leadville held in August during the height of the magnificent mountain summer. This gathering of tourists and locals is known as "Boom Days." While the gin and margaritas are being delicately sipped over in Aspen, the Jack Daniel's and Coors is being belted down in Leadville while hardy, masochistic souls engage in a grueling burro race over 13,000-foot Mosquito Pass, the highest in the country. The race starts and ends on Harrison Avenue and is no easy hike at all. The burros cannot be ridden, but must be pushed, pulled, or otherwise suckered over the twenty-mile course. And while honored guests speak at writers' conferences and proud parents watch their kids jump around in ballet tights at Aspen, people will crowd around a roaring Joy air compressor to have their ear drums pounded to jelly while some Climax miners compete to drill the fastest hole in a sacrificial boulder that has been hauled from its immemorial resting place to the center of Leadville. You may like one, or you may like the other, but there are few who will like both.

By the end of August, the highest peaks in the Sawatch Range were already dusted with snow and, although it was still summer in Leadville, the nights began taking on more and more of a chill. The snow line descended steadily as September progressed, and when the Aspen turned into a blazing red-orange, one could sense the inevitable onset of the high country winter, even though there was still good weather left. The first snow swirled through the Leadville streets in early October, not yet sticking, but letting everyone know there was a lot more where that came from. Denver and Grand Junction were still in the nineties, but the early morning October temperatures in Leadville were already nudging zero. The high country summer had suddenly come and gone, and there seemed to be a great deal of truth to the old Climax miners' saying, "Missed summer that year, was workin' days that week."

The arrival of autumn meant nothing underground as the rock was steadily hauled out of the Climax drifts. There were quite a few new faces now, and many of the hands who had helped me out when I hired out were gone, seeking their fortunes elsewhere. Pat Kelly must have figured out why he was at Climax in the first place, because he put in his last shift, brassed out for the final time, packed his things, and quietly drifted back into the nowhere he had come from. His departure changed the car pool arrangement, and I began riding with the ex-recon-cop and Dan Larkin, the English major. The political mix assured one would never fall asleep while driving, since the inevitable discussion of what was wrong with the country would begin before we left Leadville. By the last few miles of each trip, any constructive arguments had degenerated into shouting matches with names like pinko, fascist, warmonger, Bircher, and many others rattling the windows of the car. Just how those two could stay friends was beyond me, but it was probably just as well they didn't work together in the headings.

About this time a particular country and western song was making it big. It seemed purposely written for all of the transient souls that lived in the Vendome and Leadville's other old hotels and who worked in the hole. It was a Kristofferson song sung by Johnny Cash, and was played by KBRR at least once an hour. Half the guys at the mine mouthed the words throughout the shifts, words that told of waking up on Sunday morning with a splitting headache and having a beer for breakfast, of rummaging through closets in search of a half-clean shirt, and of stumbling down the hotel stairs only to sense the loneliness of a Sunday morning. It was a painfully accurate description of what went on in the morning after a Leadville Saturday night, and of the people who lived in those old hotels, people who would be there one day and gone the next. To this day, whenever I hear that song, no matter where I might be, I am once again driving up that hill for another shift at Climax.

The end of October also brought the hunting season, and as popular as guns and four-wheel-drive vehicles were among the miners, I never understood why Climax just didn't shut the mine down. The old timers who had seniority had put in for vacation time months before that would coincide with the Colorado deer and elk season. Everybody else would simply make the time, and Reasoner knew it. About a week before the season, he came around to each heading to get an idea of whom he would have for a crew on that first Monday. Black had already given him some line to account for his planned absence.

"Gonna be here Monday, right?" Reasoner asked and told me.

"Damn, Kenny," I said in a troubled voice, "I'm just not sure."

"What's the problem?" Real concern in the question.

"Got some important personal business," I lied. It was a stupid lie anyway, since we were working grave on Monday.

"You know, I just might be short a few hands on Monday. I sure could use you."

"It's pretty important, Kenny, I just don't think I'll be able to. But I'll sure try." I said that with the best conviction I could muster.

He pushed his hard hat back, drew a pack of cigarettes out of his shirt pocket, pulled one out, studied it for a while, then slowly lit it.

He exhaled the first puff of smoke, then asked, "Where you goin'?" No expression at all on his face.

What the hell. "Up past Steamboat, close to the Wyoming line."

Reasoner pursed his lips slightly and nodded his head faintly in the affirmative. "Think you'll do any good?"

"Who knows? I hear that snow last week brought a lot of 'em down out of that high stuff. Good a chance as any, I guess."

"Who you goin' with?"

"My Pard."

A weak smile. "I shoulda known. Well, hope you bring something back. How long you figure to be gone?"

"Two, maybe three days," I replied honestly.

"Well," he said with resignation, "get back as soon as you can. I can't run this goddamn place by myself."

As he turned to leave, I asked, "You goin?" I knew he had quite a reputation as a hunter himself.

Looking back over his shoulder, Reasoner said with a rare grin, "I just might."

After the first few days of the big game season, Climax got back to normal. Most of the headings were said to have been shut down on Monday for lack of miners. For the next few weeks, the allotted half hour lunch, which was usually stretched to forty-five minutes anyway, was extended still further, to nearly an hour. During this period, when one had a truly captive audience, the sagas of the great elk that fell, and even of those that didn't, were recounted in intimate detail like the ancient Indian rituals in which the adventure of the hunt would be acted out for the whole tribe. Many of the stories were embellished to a greater or lesser degree, but this was done in all innocence. The exaggeration of certain points was simply to help convey and express the excitement, drama, and humor of the moment. For example, I had managed to get myself lost for a few hours. But to hear my partner recount the story to the crew, one would have though I wandered into Canada, and the account of their actions and concern for my welfare was a little overdone. The truth of the matter was they sat in the tent sopping up my booze wondering when the hell I was going to find my way back.

By November, the country around Leadville had turned to the gray, brown, and white that would prevail for the next seven months. At first the snow is novel and really appreciated, since it allows the skiing, snowmobiling, snowshoeing, and tracking to begin. But one can only ski and run around the woods freezing for so long, and, gradually, most of the outdoor activity shifted into the bars along

Harrison Avenue, where the long siege of winter might better be held off until May. Leadville had its fair share of bars, and I doubt whether the owners ever got behind on their payments. I had always been amused by the difference between a "lounge" and a "bar" and have never been able to pinpoint that gray area where a "bar" might move up into the "lounge" league. It will suffice to say that Leadville did not have any lounges. What it did have was a number of hard-drinking bars, some of which were nearly ninety years old. Both the Silver Dollar and the Pioneer got their starts filling glasses for the silver miners, and they haven't stopped yet. Both still have the original trappings of the 1880s, the cut glass mirrors and the original bars, worn smooth by a lot of elbows. The Leadville bars were surprisingly subdued most of the time, not at all what one might expect from a bunch of miners. The conversations mostly dealt with the mine itself, and were conducted in low voices over tightly held glasses, with one eye on the door as though winter at any moment might kick the door in and freeze them all.

Black and I were assigned to a heading on the 629 Level, a ventilation level beneath 600, and were the only team down there. We had about two miles of drift to run around in, and we used our "private" motor quite often to chase down equipment at the far end of the level. Some of the equipment could have been picked up right at the shaft station. Few miners have their own level, and having one had certain advantages. Whenever a distant light came bobbing down the drift, we could be pretty sure it was Reasoner. It would take three or four minutes for that distant light to reach the heading where Black and I could often be found on our backs on a timber flat. With the exertion and the elevation, particularly on grave, it was damned easy to fall asleep in the process of taking a much-deserved break.

"Move your light, Pard," Black would say.

"Somebody coming?"

"Yeah, probably Reasoner. He's about due."

Since two beams of light were the only things a person approaching down the long straight drift could see of the heading, it would naturally look good if they were moving. The two of us would lie there until the last moment, absurdly shaking our hard hats and making sweeps across the back or the face with the lamps. The idea was that light movement would mean you were not lying on your dead ass but rather doing something related to work. This is not to say it was possible to lie around all shift. At the end of a shift, either a certain amount of work was done or it wasn't. If it were, only two people could have done it. If it weren't you'd better have a good excuse. This was the difference between "day's pay" and contracting, about which I would learn later. With the day's pay system at Climax, you did a reasonable amount of work and that was it. No heroes.

In our idle moments in the lunchroom, Black and I had adorned the backs of our yellow wet suit jackets with heavy felt marker print reading "REASONER'S RAIDERS," and, in smaller print, "The Way to Mine On 629." Finally, at the bottom was "Lunchroom Lightning," underlined by a lightning bolt, which aptly described our daily arrival at the lunchroom. Our elaborate jackets were officially condemned by the shift foreman who told us brusquely to take the wording off. We never did, and surprisingly never heard any more about it. I'm sure it embarrassed Reasoner, who was serious by nature, to have part of his crew running all over the mine where they had no business being in uniforms bearing his name.

Often it would be possible to tell from the swing of the lamp and the gait of the man under it who was approaching the heading. Usually only two people ever came down to 629, Reasoner and Ted Wiswell, the pipeman on the crew. Whenever the light got close, we could pretty well tell who it was.

"Don't look like Reasoner," Black would say.

"Nope, walks like its dead. It's that pipeman of ours. Want to break his ass?"

"May as well, we do every other day."

As pipeman, Ted, who had been at Climax in the underground for fourteen years, made regular rounds to each of the scattered headings. He lugged a canvas bag of pipe fittings, hose clamps, connectors, and everything else that was required to fix the hoses that were constantly being run over by the trains and cut in half, or to advance the service pipes to keep up with the headings. Ted was pretty good sized and appeared even bigger in his coveralls and heavy clothes. Carrying that bag, he would always remind me of Santa Claus. We both looked forward to his rounds as something to break the monotony of our exile on 629.

"I'll talk to 'im and keep him busy. You stuff the rocks in his bag this time," my partner would say. "Not too many now, about ten pounds worth."

Slowly, the light would make its way into the heading and Ted would swing the heavy bag off his shoulder. "Well, what's Reasoner's Raiders doin' today?" he'd ask. "Nothin' I suppose."

"Jee-sus Christ, I wish I had your job," Black would start off. "What a racket, nothin' to do but walk around and hide from Reasoner, take a nap here and there in some empty lunchroom, all the time gettin' paid for it. Hell, you ain't been down here in a week."

"I was down here yesterday and you know it," came the retort. "I'm down here every day. I get to every heading every shift, not like the rest of the pipemen in this mine. You think it's easy tryin' to keep up with you clowns the way you cut them damn hoses in half every shift? If you'd take a minute to pick the goddamn things up instead of runnin' 'em over, but no, you're too damn busy for that, cause you're miners."

"Hey, Ted," I'd ask, "did you ever do any mining? I mean real mining, not just carrying that bag around." I knew what the reaction would be.

"I done more minin' in the first couple of months I hired out here than the two of you done in your lives, and that's the truth," he'd say with the gravest sincerity. "This was a workin' mine back then, not like it is now when they hire every dumb sonuvabitch that walks in the front door."

Ted had hired out in 1956, when the company town was still in existence, and I always enjoyed listening to his stories about life in Climax when the hotel, houses, company store, hospital, and recreation facilities made a self-sufficient community on top of the Rockies. In '56, Ted would tell us, a man could come up to Climax and hire out as a miner for eighteen dollars a shift, and, if he chose to live at the company hotel, pay only twenty-one dollars per week for room and board. They were, Ted always said, the "good ol' days." The company town was phased out in the early 1960s, and most of the buildings moved down the hill to Leadville, leaving only the U.S. Post Office, the highest in the country, and a misleading town dot on highway maps.

While Ted would be talking about the days gone by, I would be busy nonchalantly stuffing his bag full of rocks. "Well, I can't sit here on my ass all shift the way you miners do," was his standard closing comment. He'd pick up his bag, swing it over his shoulder, and trudge off into the darkness.

"Did you load him up?" Black would ask.

"About ten pounds worth."

The two of us would roll back over on the timber flat laughing like hell at the thought of how angry Ted would be when he got to his next stop and empied that bag. Twenty pounds of fittings and ten pounds of muck. It was the little things like that which made the long hours move.

One thing about mining, I learned, was that things could go wrong instantly, ruining what was to have been a good shift. At the end of one of those uneventful shifts, Reasoner, who, without showing it was fairly pleased with our efforts, told Black and me to bring our motor up the incline to 600 Level for maintenance. The 600 and 629 Levels, besides being connected by a shaft, were also connected by a railroad that ran up an inclined drift. As soon as Reasoner had left the heading we packed our lunchpails and dulled drill bits on the little five-ton motor, swung the trolleypole around so it would trail behind us, and began the nearly mile-long trip. We had to take our sweet time since it wouldn't do to show up at the shaft station on 600 too early. The two of us sat perched on the little red motor, Black controlling the throttle with his feet, sipping the rest of our coffee, clattering and swaying down the bumpy track under the intermittent flashes of the trolley runner. Halfway up the gradual incline was a level area with a rail switch point leading off into a small siding. The pumps had apparently failed and the switch point was flooded with about two feet of water.

"Pump's out again," Black noted.

"Yup," I replied. Real concern. End of shift anyway. It'll give the guys coming on something to do. "We'll tell Reasoner."

We took the motor slowly through the water and ran into our first trouble when we passed over the switch. It wasn't thrown completely and we split it, the front wheels of the motor taking the rails up the incline, but the back wheels somehow following the rails onto the siding. As slowly as we were going, we were able to stop the motor when we felt it lurch, preventing derailing.

"Split it," my partner said, his voice resonating in the sudden silence.

We screwed the cups back on the thermoses, repacked our lunchpails, and checked our watches. It was no big deal to back the motor over the switch, then get a bar or something and lever the switch blades together. We shouldn't even have to get our feet wet. One of the first things every new hand is taught to not do is backpole a motor, that is, to be too lazy to swing the pole around so it would follow safely at a trailing angle. Since this required climbing up on the motor and physically swinging the heavy pole, it was generally regarded as a nuisance, especially if the distance to be traveled in reverse was short.

"Now I'll tell you what'll happen if you backpole these motors," Wizen had told us nearly a year before when I was a new hand. "That damn pole will hit a burr on the line and she'll snap in two. And when she does, if you're in the way, she'll knock you silly. Now I've run my last motor, so it doesn't mean anything to me, but for your sake, since you are the guys who are going to be running them, please take the time to swing those poles around."

I looked at Black, he looked at me, and in silent agreement he kicked the motor into reverse with his foot. Everybody in the mine backpoled at one time or another, and a lot more than the ten feet we had to go. All of a sudden, everything that could have happened did. The motor suddenly lifted on the rails and fell, derailed. But in the instant before it fell, the trolley pole jammed against the line, shattering with the crack of a heavy rifle and throwing splinters and the broken top of the pole over the top of the motor. It broke, unfortunately, a little too late, for the thrust on the trolley itself was enough to part the uninsulated four-hundred-volt copper line. That separated with a second monumental crack and a white flash like a bolt of lightning. Before we had any time to react at all, the loose trolley line snaked over our heads and flew against an air and a water main on the side of the drift. This caused a second bolt of lightning followed by an enormous shower of red sparks as the trolley line burned through both pipes and released a column of water and the explosive scream of high-pressure air. In less than three seconds, the quiet, gloomy drift had turned into a combination of Judgment Day, Armageddon, hell, and the engine room of the Titanic as it went down. If Reasoner's Raiders could have run any faster, we would have been running on top of the water.

When we reached the dry ground of the incline, we were both soaked from splashing and stumbling around in the thigh-deep icy water. "You okay, Pard?" my wide-eyed partner asked.

"Just wonderful," I panted.

"Sonuvabitch," he said between breaths, "that was enough to scare the shit out of a wooden Indian."

A short way up the incline we found a trolley switch to stop the arcing, then shut an air valve to cut off the scream of the escaping air. We looked back to the switch point, through the smoke, and saw the motor leaning at an angle in the deepening pool of water. Only the gusher from the burned-out water pipe was still making noise.

"Goddamn, is Reasoner going to love this. It'll take 'em all next shift to clear the wreckage," Black remarked. We walked the remaining way up the incline and waited for the cage at the station.

"Where's the motor?" Reasoner asked.

"Down at the incline pump station, Kenny," Black said with an air of nonchalance. "Some ass on the other shift screwed that switch up and we had a little trouble."

"What kind of trouble?"

"She's on the ground."

Reasoner grimmaced.

"There's a little water down there, too."

"How much?"

"Oh, about two feet."

Two feet high and rising, I thought, silently admiring my partner's subtle approach to informing Reasoner of the disaster we had caused by a flagrant breach of safety regulations.

"Oh yeah, something's wrong with the trolley down there too. No power," Black went on, dutifully informing our shifter.

"Okay, I'll get 'em to send a pair of hands down there next shift," Reasoner said.

A pair of hands, I thought, along with some electricians, a repair crew, and a couple of pipemen. Amazingly enough, we never heard any more of it. I could imagine what the crew coming on would have to say about the asses on the other shift. No harm had been done that couldn't be repaired, and, although Black and I laughed about the episode for many shifts thereafter, the point of it was indelibly impressed in my mind. One simply never knew where or when trouble, serious trouble, could start in an underground mine. Granted, we had brought all that upon ourselves by what seemed at the time a very minor infraction of safety rules, but even relying on all your experience and meticulously adhering to the safety regulations at all times was no guarantee that you would be immune to sudden injury or death underground. An individual miner's welfare depended not only on his own actions, but upon the actions and judgement of many other people over whom he had no control and in whom he had to place his trust. That was something that would be proven to me several years later in another place, at another mine.

As Christmas approached, the idea of five full months more of the Leadville winter began weighing heavily on my mind and, more and more, I could hear a distant rustling of a warm tropic breeze through the palm leaves. Thinking back, I have never been able to pinpoint the actual day I made the decision. I thought often of Pat Kelly and his "when you know enough to leave, you'll know why you were here." I finally told Reasoner I would quit just before Christmas.

"What's the problem?" he asked.

"My fingers are worn down to the bone, besides I got a shifter I can't put up with any more." I said jokingly, but in a disgusted tone of voice. "He expects me to keep carrying the whole crew."

His face broke into a rare grin and he asked, "Seriously, why are you leaving?"

I told him I wasn't putting up with five more months of the Leadville winter and that was it.

"Winters always come to an end," he reminded me.

"Yeah, and they always start again. And up here it's too damn quick."

"Well, it's up to you, but I think you ought to stick around. You turned into a good hand." At first I thought that was the company man talking, because Reasoner, the man, rarely handed out compliments or pats on the back.

"Thanks," I said, "but I don't think so."

"Well, at least you ought to stick around till the twenty-seventh."

"What's doing on the twenty-seventh?"

"You'll get two days of holiday pay for nothin', that's what's doin'." I knew that wasn't the company man talking.

Two days of pay for nothing from Climax sounded too good to pass up, so I hung around and finally brassed out and hung up my lamp for the last time a few days after Christmas. I had spent the last few shifts highgrading with only mediocre results, but enjoyed walking around to all the headings on three levels to say good-bye to the miners, all of whom I now knew quite well.

I made the rounds of the Christmas parties in Leadville, and never before had I seen such sincere holiday spirit. I don't know whether it was the high country winter, the swirling snow and bitter cold, that made one appreciate being indoors with friends, but the music, the log fires, the booze, and the spoken words had never seemed warmer. A lot of that, I'm sure, was based upon the mutual sharing and enduring that took place at 11,300 feet in those dark, dreary drifts that made up the Climax mine.

One of the last stops I made in Leadville was at my partner's place. Black and I had long engaged in the materially meaningless but emotionally gratifying ceremonial exchange of bottles of Jack Daniel's on festive occasions, and this, of course, would be no exception. However, since the horseplay of the mine carried over to the surface, I couldn't resist a final effort. When I bought the gift bottle of blackjack, I also sought out what I deemed to be an appropriate wine.

"Give me the cheapest bottle of wine you got," I told the proprietor of one of Leadville's numerous liquor stores.

"We don't sell any cheap wine," the old man replied with an offended look. Like, what kind of place do you think I run? He did mention that he had several wines, however, which might be what I was looking for.

I carefully selected a bottle of some clear stuff called Silver Satin, which was a steal at sixty-nine cents, and wrapped it in a brown bag. When I got over to Black's place and he saw me walking in with a wrapped bottle, he automatically reached for the gift carton bottle that was to convey his farewell tidings to me.

"Merry Christmas, Pard," I said, my voice just brimming with the holiday spirit. "Here's the best to you, and you're worth every bit of it."

"Why, same to you, Pard," he said amid the handshaking and exchanging of bottles. His voice was a little subdued, probably since this would be the last time we would see each other. Withdrawing his gift from the paper bag, he found not the expected bottle of Jack Daniel's, but a bottle of disreputable wine with the winos' special price clearly marked on the label. His lips set in a firm line as he fought to suppress a sarcastic grin. He probably wished dearly that he had never given me the real thing, and that he hadn't been foolish enough to assume I would consider this final exchange of gifts an inviolate and sacrosanct ceremony. But he kept his cool and just said, "Why, Pard, you can bet I'll put this with my special collection. I just wish I could thank you right."

In the course of the evening I forgot to give him his intended bottle, the real one, that was in the trunk of the car. In fact, it was the next day and I was halfway to Mexico before I realized it. Well, not much I can do about it now, I thought, good thing my Pard had a sense a humor.

Years later, I was to learn that a miner's sense of humor might even be surpassed by his sharpness of memory.

CHAPTER III

ARIZONA COPPER

There is a great deal of truth to the idea that human memory acts as a filter over the years, recalling the good moments with far greater clarity than it does the bad moments. I don't know whether this is a subconscious screening process that tends to eliminate or at least diminish the recollections of experiences that once caused anxiety or fear, or whether the enjoying and rewarding moments are just normally impressed more firmly in the memory. At any rate, after two years Climax had become a pleasant, faraway montage of the good shifts, the laughing and joking, the friends, all the little humorous events that helped to pass the time, the alpine summer, the clear mountain air, and another way of life I had assumed easily. Yet the memory somehow seemed detached from reality.

In the spring of 1973, reality was a Las Vegas apartment about a block off the Strip. Once again the time had come to seek gainful employment. In other words, I needed some money. Glancing through the employment ads of the southwestern newspapers, I mused over the endless varieties of jobs supposedly available, and wondered how the unemployment levels could be so high. Again I thought of the personnel managers who placed all those thousands of ads, and of all the thousands of eager applicants who would answer them. Only a few of them would possess the attributes, not necessarily experience or education, that would render them suitable for the particular position.

Rather than dwell any longer on the inevitable, I took a tour of the blackjack tables and on the way back, picked up a copy of the *Los Angeles Times*. There it was again. The big block ad: "Underground Miners Wanted Immediately for Work in New Copper Mine Near Casa Grande, Arizona. Experience Necessary. Write or call for application. Hecla Mining Company." I quickly turned the page and skimmed over the rest of the listings but then, as I somehow knew I would, found myself staring at the Hecla ad. I thought back to Climax and how damnably easy it had been to hire out. Bet it's the same way here, I thought, just show up and start work. I didn't really want to go underground again, but the two years away from the hole had mellowed the deafening roar of the drills and softened the deadly fall of the slabs. The next morning I phoned Casa Grande for an application.

All Arizona towns seem to sprawl out into the desert, rather than grow upward, and Casa Grande was no exception. About midway between Phoenix and Tucson, it materialized out of the desert next to a ridge of low, barren mountains where there shouldn't have been a town at all. There was a lot of activity for a small town. Motel rooms were at a premium because of the influx of workers at the nearby ASARCO open-pit copper operation and because of the rapidly expanding Hecla Lakeshore Project. The "near Casa Grande" that was mentioned in the ad was mostly wishful thinking, because the mine was thirty-five miles southwest of town in the hottest, most inhospitable desert I had ever seen.

The Hecla personnel office was in a makeshift trailer and the manager was a heavyset, middle-aged man whose air of impatience told of being required to interview too many people in too few hours and to shuffle a correspondingly large amount of paper. "Climax," he murmured over the hum of the air conditioner, ignoring the rest of the information on the form. "What did you do at Climax?"

"Drift mining."

"Drilling? Blasting?" he asked.

"Yeah. And timbering, mucking. Regular drift development work, all on rail."

"Well, we haven't any rail here, all diesel. Let's see," he said, studying the application, "how long were you up there? Yeah, almost a year. Why did you quit?"

"Wanted to do some traveling."

"The simplicity of that explanation apparently satisfied him, for he nodded his head slightly in a positive way, then looked up and with a knowing grin asked, "If I check this out with Climax, they're going to confirm it, right?"

I smiled back and told him they would. I guess he had been interviewing a lot of self-proclaimed miners with imaginary experience.

"Well," he finished after a pause, "I can give you a job as an underground miner. Want it?"

"When do I start?"

"Tomorrow."

I filled out the required forms, took a physical examination in the afternoon, and showed up for work the next day. There were no wasted words or motions in that hiring process. It was purely an immediate functional arrangement between a miner and a mining company. All the company cared about was that the man could operate the equipment with a minimum of supervision and take care of himself underground. They had neither the time nor the inclination that Climax had to take on an inexperienced new hand. Aside from a few questions about where the miner had gotten his experience, where he had been and what he had done meant nothing.

I was told to report to a safety lecture room on the surface for a day of orientation, safety talks, issuing of gear, and general familiarization with mine procedures. When I walked in, there was only one other man seated. I looked around the room, which was decorated with instructional blowups of mine equipment and first aid diagrams, and asked, "You just starting today?"

"Yeah. If you are too, you're in the right place. Sit down," he said, kicking a folding chair out with his foot. He was about the same age I was, but with the weathered features and calloused hands that told of many more years in the mines than my one. "Where you from?"

"Mined up at Climax for a year," I answered, kind of enjoying the fact that this wasn't going to be a "new hand" experience all over again. "You?"

"Come out of Grants."

In the few minutes we had to talk before a Hecla safety man showed up, I learned his name was Billy Shelton and that he had been working in the mines for nearly eight years, all his working life. Most of that time was spent in the uranium mines of Grants, New Mexico, which had had just tramped the previous week. There was a lot of talk up there, he said, about Hecla hiring in Arizona, and he decided to find out firsthand what it was all about. His family would be leaving Grants in a day or two to meet him here in Casa Grande.

The instructor introduced himself and gave us a little of his personal history, including the fact that it was a broken back suffered in an underground accident that led to his present position in the safety office. His talk was very informal and concentrated not on falling rock as the biggest hazard, but rather on the possibility of an underground fire. Unlike Climax, which used electric rail power, Hecla relied on diesel power in the underground. This meant there would be large quantities of fuel, hydraulic fluid, motor oils, and various other flammable lubricants underground at all times. All of the diesel equipment was equipped with remote-control dry chemical fire systems, portable extinguishers, and self-rescuers, which were individual emergency breathing devices for use in underground fires. The real danger in underground mine fires was not the physical danger of the flames or smoke, but the increasing concentration of highly toxic carbon monoxide, an odorless and tasteless gas, a product of combustion which could accumulate rapidly in the confines of underground drifts.

The classic example of what an underground fire could do was shown only a year earlier at the Sunshine Mine in Kellogg, Idaho. Ninety-one men of the shift total of 173 had been killed by carbon monoxide poisoning in the second greatest tragedy ever to strike a United States hardrock mine. As a result of the Sunshine disaster, the

U.S. Bureau of Mines was taking steps to make it mandatory for every underground worker to carry a self-rescuer with him. At the time I hired out at Hecla, the hundreds of self-rescuers that would be necessary for everyone working underground were on order.

We were shown the operation of the little can-shaped devices, and the instructor went on to explain the stench warning system in use at Hecla that was to be activated in case of an underground fire. A gas with a disagreeable and distinctive odor similar to garlic would be released in the compressed air system as a warning to all underground workers that a fire was in progress and that a carbon monoxide danger might exist. We drew our gear and brass and spent the rest of the day telling lies and coming up with an endless list of names to see who knew whom in what mine where. It was here that I learned that hard-rock miners are locked into a limited circuit and that their names and reputations lingered long after their departure and often preceded their arrival at new mines. At the end of the day, Billy and I were assigned to different mine crews on the same shift. As soon as we got settled in Casa Grande, we agreed, we'd buy each other a couple of beers.

Access to the underground levels of the Lakeshore mine was provided by twin parallel declined tunnels, which started at the portal elevation of 1,915 feet and drove downward at an angle of minus fifteen degrees. The North decline, a large drift nineteen feet wide by fifteen feet high, was floored with a standard-gauge railroad on which men, supplies, and muck were transported on a large twenty-four-ton skip. The skip was little more than a multi-purpose rail-mounted cable car with bottom dump doors for unloading muck on the surface, and with a removable rack which provided seats for the underground crews at shift changes. The South decline, one hundred feet away from the North, was about fifteen feet wide and fourteen feet high and was maintained as a roadway to permit the rubber-tired diesel equipment to move in and out of the mine. Once the mine went into production, ore extraction would be accomplished by a sub-level caving system from beneath the large, complex ore body. The mine was now only in the development stage, and the general conditions, I would soon learn, would be far different from those of an established production mine. Both declines, when completed, would be about 7,000 feet long and terminate nearly 1,900 feet beneath the Arizona desert. They were then about three hundred feet short of that point. The declines had become a priority job since much of the work on the levels hinged on their completion.

The skip was boarded on a surface steel and concrete ramp and descended rapidly, clattering and swaying along the rails much like a

Rail skip bringing the graveyard crew out of the decline at Hecla's Lakeshore Project. (ASARCO Incorporated)

downhill subway ride. I remember quite clearly my first impression of Hecla—the smell. I had never realized that all mines must have that same characteristic odor of dampness and dust, the smell of the rock itself. As soon as that familiar smell hit my nose, the eleven months at Climax flashed through my mind with alarming lucidity, and I was suddenly very unsure about whether I really wanted to go through all this again. The skip flashed past pumps and pipes, stations and crosscuts, all lit by the string of bare electric light bulbs running down the center of the back, and three minutes later it eased to a stop under the harsh glare of mercury vapor lamps at the 500 Level station.

Just prior to arrival at the station, where the tracks temporarily ended, the miners had fumbled around putting on ear protectors, and now it was very apparent why. Mounted overhead was an enormous ventilation fan, the scream of which was so loud it was painful to the ears. Any conversation at the station was limited to shouting directly into a person's ear. Since I was assigned to the decline crew, I was led to a diesel-powered jeep that took us to the lowest level of the mine, to the faces of the declines themselves.

The next day at the start of the shift, I saw Billy Shelton. "Where they got you workin'?" he asked.

"Decline headings. You?"

"A raise on 500."

"What do you think of this place?"

He shrugged and smiled, "It's a mine."

That's for sure, I thought to myself, glad that I had wound up in the declines and not in a raise. At Climax, I had spent only about a week driving a raise and wasn't overly enthused about it. A raise is a vertical tunnel that is being driven upward from below. The same vertical working, being constructed from the surface or from a higher point in the underground, is referred to as a shaft, and its construction is known as sinking. So, in the nomenclature of the mines, you drive a raise and sink a shaft. Driving a raise means the miners must work directly beneath the rock they are drilling and blasting at all times. Rock falls in a raise can be particularly dangerous since, while working either on a mechanical climber or a simple timber platform, a miner is trapped like the proverbial rat.

Billy and I talked until the roar of the vent fan at 500 station made conversation impossible. He went his way and I got into a jeep for the ride down to the decline headings. Little did I know that that would be the last time I would ever see Billy Shelton.

About an hour after the shift had started, the headlights of a jeep bounced down the rutted floor of the decline. The big hulking man who got out was Luke Byrd, one of the shift bosses, whose sagging bloodhound jowls gave him a perpetually sad look. I had liked Luke the day I met him, and soon learned that his slowness of speech and movement was born of caution and deliberation, and not of any laziness or lack of drive. We hadn't expected to see him so soon after the start of the shift, since most of the Hecla shift bosses, after lining out their crews, exercised their management privileges by retiring to the 500 mechanics bays to fill out their time books and stare at the walls in relative peace and quiet.

"Fellas," he started out in a somber tone of voice, "we just had a bad accident up on 500. Man caught a slab up in that No. 4 raise and got hurt bad. His two partners got banged up a little, too, but they're okay."

"Well, how is he?" one of the miners asked.

"He was still breathin,' just breathing' I guess, when they hauled him out of there."

"Who was it?"

"I forget his name. New man, only been here a few days." Then Luke nodded at me and said, "Hired on with you, I think."

"His name Shelton?" I asked.

"Yeah, yeah it was. Shelton. Think he came in from Grants. You know him?"

"No," I said, "just hired on with him."

Luke paused, then went on. "Okay, fellas, we had enough trouble down here already today, we don't need nobody else hurt. Just do what you can and be careful." Although I didn't know it at the time, that was one of the few times anyone would tell me to take my time doing anything.

About two hours later, three hours into the shift, the word came down to the declines that Billy Shelton was dead. Because of the size of the declines, we worked in groups of three miners, and the three of us just looked at each other and said nothing. I remembered Billy's telling me only yesterday that he had tramped Grants because he was vaguely dissatisfied, no real reason, just that if there are better mines, you won't know about them until you work for them. I wondered what his widow, who was on her way from Grants to Casa Grande at that very moment to join her husband, would think when she learned that she no longer had a husband. Also, I visualized the imaginary balance I had mentally concocted at Climax, with the paycheck on one side and that big slab of rock on the other. I never really knew where my own needle rested since I viewed the mines as a temporary thing and never allowed the pans to stop rocking. At the moment, the slab of rock was winning out over the paycheck. I doubt very much whether Billy Shelton ever thought about things like that. In his hard hat and lamp and with a string of drill bits slung over his shoulder, his face reflected nothing but total acceptance of the fact that he had been a miner all his life, and probably always would be. I imagine he had rationalized the hazards of underground mining fully, and compared them to any other everday event or occurrence of life. They were unavoidable, something that he had had to face, and that was that. So simple.

Billy and his two partners had climbed the wooden ladder into the timbered raise at the start of the shift and two of them had taken turns barring down the back. When the back had been pronounced sound, they had run some flexible air vent tubing to the top of the raise, then had sat back and waited for their shifter to line out their work for the day. While his two partners had chosen seats along the edge of the raise, Billy had sat on a timber across the center. Hindsight is the best sight in anything, particularly when it comes to another miner's accident. Still, any experienced miner can tell you that when killing time, do it along the rib, and not directly beneath unsupported back. He would also tell you that the introduction of a fresh air flow to a rock surface may produce cracking and slabbing, however rarely. Perhaps his seat was to have been temporary, but that is no longer important now, for a slab of that cold, hard rock about five feet long had let go from the middle of the back and struck Billy squarely. He had been taken down the raise as quickly as possible, given heart massage and mouth-to-mouth resuscitation, brought to the surface to a waiting ambulance, and sped thirty-five miles through the desert to Casa Grande. He was pronounced dead in the emergency room a short time later.

It's a funny thing among miners how the close calls and the more serious underground accidents involving injury or death are recalled so differently. The close calls are recounted in intimate detail with the number of inches that made the difference, the words spoken under duress at the time, the personalities of the miners involved, and the painstakingly accurate descriptions of the accident area. These stories are passed from heading to heading, from shift to shift, and the really good ones from mine to mine. They became something akin to legends. The fatalities and the severe injuries, however, are referred to only in the vaguest of generalities. Names are frequently omitted, and the details, which somehow seemed irrelevant anyway, are lost in time. Billy Shelton's death was destined to become just "that guy who got slabbed up in No. 4 raise a couple of years ago." The reason is obvious. The selective recollection is a conscious mechanism to offset the fear and awareness that anyone can run out of luck at any given moment in any mine. And every miner knows that his accident, when it comes, will be one of the close calls, one of the legends to be recounted at the lunchroom tables, in the headings during the breaks, and in the dry during shift changes.

A collection was taken up and the proceeds of more than $800 donated to Billy's widow. A xeroxed copy of the check and a brief note of condolence from the Hecla Mining Company and the underground crews were tacked up at the bottom of the bulletin board in the mine dry. A few days later, a short thank-you note was tacked up alongside. In a few weeks, both became dusty and yellowed and were replaced by crew schedules.

The mechanical workhorse of the underground was a huge beast known as the Wagner ST-8 Scooptram. It was, in essence, a front-end loader that had been specially adapted for underground use. The overall length of the ST-8s, from the end of the engine compartment to the tip of the gaping eight-cubic-yard front bucket, was a respectable thirty-nine feet. The machine rolled on four enormous six-foot-high tires and had a maximum height of 6½ feet. The top of the unit was completely flat with no protruding parts save the head and shoulders of the operator, thus allowing it to creep beneath most of the maze of pipes and vent tubing that cluttered the drifts. The loaders, as they were called, had no suspension system, and steering was articulating. That is, the vehicle was hinged in the middle and could twist at the waist, much like the segmented body of an ant, enabling the big machine to negotiate the tight turns in the underground drifts. A powerful hydraulic system ran both the steering pistons and the versatile bucket, which could be raised and lowered as well as tilted forward and back.

A ten-cylinder diesel engine developed 250 horsepower and provided the motive power for the ST-8s as well as their air compressors and hydraulic pumps. This wonderful machine held more than 160 gallons of diesel fuel and, as if that weren't enough, about 140 gallons of hydraulic fluid. When I learned of these specs, I realized why there was talk of fire danger when I hired on. Each ST-8 was equipped with a thirty-pound dry chemical extinguisher, the jets of which were aimed at likely areas of the engine. The jets could be activated by slamming a handy red button on the instrument panel. A smaller, portable unit was mounted near the operator's leg.

Hecla had a small fleet of ST-8s running rampant around the drifts and declines, terrorizing miners on foot while performing or assisting in virtually all of the tasks required in the development of the mine. The loaders would haul muck from the headings to the skip

A small load-haul-dump (LHD) unit in operation. Note the narrow drift that leaves little safe space for miners on foot. (ASARCO Incorporated)

loading point on the 500 station, haul supplies from the station to various points throughout the underground, and the surprisingly manipulative bucket would serve as a work platform for miners. When new, the loaders had been a bright yellow, but a few weeks underground had reduced them all to a uniform fleet color of grease, dust, and muck. There were also a couple of ST-2Bs running around, a far smaller look-alike, doing their best to keep out of the way of the charging steel buckets of the massive ST-8s.

The loaders were relatively simple machines to operate. The operators sat on the left side of the unit facing not toward the bucket, but toward the middle. Running one of these machines on the surface beneath the warm Arizona sun, once one had gotten used to the size and the articulating steering, was quite safe and simple. Underground, however, was a different ball game. The ST-8s were fully eight feet wide, which, in the narrower drifts, didn't leave much room to clear the high-pressure air pipes or the high-voltage cables. Often clearance was just a matter of inches. In the underground darkness or on the slippery mud of the drift floors, a slight error in judgment on the part of an operator could undo in a few brief moments of tearing and crushing what had taken several shifts to put together. To complicate matters, the mechanics who more or less maintained these machines would often perform deeds of mechanical trickery. Among their favorites was to replace the steering gears in reverse, thus making the operators think left to make the thirty-nine-foot loader turn right. They wouldn't do this to all the ST-8s, of course, but to just about half, so that each unit would be a mystery until you got on it and found out by trial and error which way the thing steered. After half a shift on one unit, when an operator had to change machines, usually because of mechanical failure of some kind, the hesitation and doubt that came with reversed steering only increased the possibility of error and thus the probability of trouble.

If the ST-8s were tricky to handle on the underground levels, they were just plain dangerous on the declines. The declines, which dipped at an angle of minus fifteen degrees, were difficult to negotiate because of the lack of a normal horizon. If a miner spent enough time in the declines, he would be so mentally accustomed to the angle that at the end of a shift he would swear he was on a level. And when he finally did get to a level, he would find himself leaning forward at an idiotic fifteen-degree angle. Perhaps the best way to describe the actual descent would be to describe a highway grade. On mountain roads, warning signs caution trucks to use low gear on grades exceeding five or six percent. The Hecla declines were equivalent to a twenty-four percent highway grade, far steeper than any normal road grade.

Keeping this angle in mind, consider now that an empty ST-8 weighs twenty-eight tons. With the bucket full of muck, the total weight neared forty tons. And with a bucket full of concrete which they often carried, the weight surpassed even forty tons. With the combination of weight and grade, the wheel air brakes were often incapable of stopping the ST-8s. Lowest gear was all that would enable an operator to control his speed. From a dead stop, if the unit were not in gear, just releasing the brakes would produce frightening acceleration over a very short distance. Safety regulations required that the ST-8s use the declines only when the bucket was facing downhill, so that in the event of a mechanical failure and a runaway machine, the operator could attempt to steer the bucket into the rib and, with luck, wedge or jackknife the loader into the drift, stopping the uncontrolled descent. This is certainly a sloppy way to stop the unit, since damage to equipment and timber in the rib can be considerable, but it may be the only way. If allowed to run free, the ST-8 will accelerate to nearly one hundred miles per hour and, since it has no suspension, will bounce wildly off the ribs and probably even the back, in the manner of a small gauge slug being fired down the bore of a large-gauge shotgun. Just such an event happened in 1970 on the declines. The operator either bailed out in panic or was thrown off his loader and killed, with the runaway machine narrowly missing others before it demolished itself. Either way, staying with the unit or getting off, your chances are about the same to make it to miners' heaven prematurely. At runaway speeds, a loader can make it from the portal to the rock face 7,000 feet away in much less than one frantic minute. If there are crews down there, they get retirement tickets too. Mechanical failure in the ST-8s on the declines was always a gnawing discomfort in the minds of both the operators and the decline crews.

After only a few weeks at Hecla, someone decided that I should be trained as a loader operator, confirming my suspicion that it would only be a matter of time. I spent half a shift driving one of the ST-8s around the desert, dodging saguaros and running up and down hills. By lunch I was pronounced a bona fide operator, immediately fit for decline duty. After that, most shifts would start with a shift boss pointing to a waiting loader and saying, "Pard, take that one and . . ."

It was only May when Casa Grande had its first one-hundred-degree day. For some reason, the local media marks this dubious occasion as though it is some cause for joy and thanksgiving. The newspapers and radio stations all make excessive mention of the fact, much as Florida duly notes the beginning of the hurricane season. In reality, all that first one-hundred-degree day meant was that there would be damn few days between then and late September when it wouldn't reach that celebrated level. June was hotter than the gates of hell, with temperatures around 105° every day. I was already beginning to wonder how I was going to make it to September when the daytime temperatures would once again drop into the cool nineties.

The wet heat of the underground was far worse than the dry desert heat of the surface. The ambient rock temperature through which the drifts were driven varied, but it would often be as high as 120°. The heat would radiate from a fresh muck pile as an almost tangible, heavy force that just drained a man's energy and drive. Cooling the underground and providing ventilation was effected by blowing surface air down vent shafts with huge fans. You can imagine how much cooling was done on the days when the surface temperature hovered around 110°. Since the mine accumulated considerable water, the resulting humidity made for some of the worst working conditions I had ever imagined.

To top it all off, in 1973 the Lakeshore mine was under development. That means the final pattern of drifts, which would provide maximum efficiency of ventilation and haulage when completed, was only under construction. Most everything underground was of a temporary or makeshift nature and that, unfortunately, included the ventilation. There is no comparison between a mine that has reached production, such as Climax, and one that is still under development. Climax enjoyed fixed quotas of production and knew pretty well just what should be accomplished with a certain expenditure of man-hours of effort. The more routine the underground operation becomes the greater the overall efficiency and, most importantly, the greater the safety. It was slowly occurring to me that there would be little similarity in any respect between what I had known at Climax and what I would see at Hecla's Lakeshore Project.

An established producing mine relying on efficiency, not speed can afford the luxury of indulging in and enforcing an effective safety program, which, in the long run, will only work to its advantage by reducing lost man-hours and wrecked equipment. A developing mine is a far different story. If anything can possibly go wrong, it will. Every step of the development, even though it will have been blueprinted to death, is still an experimental effort. Geology is, at best, an inexact science. Core drilling is expensive, and is ordered only in locations thought absolutely necessary. Even with the two combined, the exact nature of a fault will only be known when the miners drive into it, not before. Situations and conditions are constantly changing in a developing mine, sometimes from shift to shift, and since efficient operations are possible only when some semblance of stability has been achieved, there is no practical efficiency. It is speed, with the resultant

Herschel Morton during a lunch break at Hecla's Lakeshore Project. (Stephen M. Voynick)

Dan "Shag" Harger. The intense heat of the mine is causing him to sweat. Heat is one of Arizona copper miners' worst enemies. (Stephen M. Voynick)

confusion and increased injury rate, that reigns supreme in a developing mine.

The Lakeshore Mine was a joint venture between El Paso Natural Gas and the Hecla Mining Company. Hundreds of millions of dollars had been committed to the development of the mine and the construction of the surface mill facilities. When I hired on in 1973, the project was about two years behind schedule. Planning a mine in the comfort of an air-conditioned drafting room is, I imagine, a fairly routine matter. Arriving at a supposedly realistic time estimate for completion probably involves merely the use of a few formulas with a contingency factor added on. I wonder whether there has ever been a mine that was developed according to the initial projection. One of the problems Hecla encountered, for example, was driving the North decline through a fault where 150° water gushed out in torrents. I doubt very much whether this unfortunate occurrence was included on the blueprints. A great deal of time was lost when the entire fault area had to be grouted off.

In the oak-paneled conference rooms of the corporate headquarters, far removed from the sweat and darkness of the underground drifts, little attention is given to the details of why their multi-million dollar project is two years behind schedule. The simple fact is that it is behind, and I am sure enormous pressure can be brought to bear on the resident manager and his staff to get the job back on the track. Considering that twenty-year careers are at stake, it is small wonder that the whip cracks right down the line, for the mine managers are only too aware that they may wake up one morning out in the cold and find someone else who can drive harder sitting in his office.

Such pressure can put the lowest level of mine management, the shift boss, in an awkward position, since he personally knows many of his miners and understands quite well just what the problems and holdups are. How the shifter reacts to the pressures from his supervisors sets the trend for safety and work done. I would dread those shifts when my shifter would walk up to me in the dry with a wide, irrational look in his eyes. I knew at once that in his pre-shift briefing upper management had put the fear of God Almighty into him again.

"Pard," he'd say, "we got us a lot to get done today. We got to load and shoot, get 'er mucked out, and stand a set of steel. Got to."

I'd shake my head, knowing very well how much we would get done. "Well, we may start with the steel, but . . ."

"No buts, Pard, not today. They," he'd nod his head back over his shoulder toward the dry offices and a reverent tone would creep into his voice, "want it done for sure. They're already figurin' on marking out the crosscut next shift, so we gotta get 'er done. Or else."

"Or else what?" I couldn't resist it.

He'd look at me through ever-widening eyes, as though I had committed blasphemy in the basilica. "Or else . . . or else," he'd sputter, "well, you'll find out what else."

I knew very well what "or else" was. Or else he'd get dragged on the carpet.

One of the reasons I liked Luke Byrd was that I knew exactly how he would react to the same situation. He'd climb out of his jeep and move his big frame slowly into the heading, taking a long slow look around at what was being done.

"Fellas," he'd say, as slowly as ever, "why don't you quit what you're doin' and come over here for a minute." We'd all knock off and park ourselves on the closest pile of muck or timber.

"What'd those bastards tell you about us, Luke?" someone would ask.

"They told me none of you was worth a damn."

"Hell, we could've told you that down here and saved you the trouble."

Luke would smile, then get down to business. "Oh," he'd say almost in a moan, "they think we can get this timbered up 'fore quittin' time."

"What the hell we goin' to do the second half of the shift?" some- one else would ask, his voice as serious as ever.

Luke would smile again. "They kinda think they can send some engineers down here next shift to mark out the crosscut. It would be good if we could get 'er timbered up, but if anybody's goin' to get hurt, well then we'll just give 'em what we can." After a pause, "Think you got a chance?"

"Chance it'll snow down here this shift, too, Luke."

He would rarely wait for a concensus. "You know what I mean, fellas," he'd say, getting to his feet. "Just be careful." More often than not, the timbering would somehow be finished, barring a mechanical or supply hangup. Luke Byrd, without pushing, had a way of making you want to do things for him.

The mine crews on the shift were fairly small, about ten men each, and often we worked side by side, especially when some rush project was ordered. There was lots of time for conversation, especially during the lunches when everybody would lie down on a piece of dry lagging and either talk or just let the heat lull him to sleep. One of the guys on the other crew with whom I got along well was Terry Udall, who had tramped the mines in Kellogg, Idaho and had hired on at

Hecla about two months before I had. The miners had given him the nickname Blue because of his close resemblance to the character of the same name on the High Chapparal television series. Blue was as easygoing a person as I have ever met, and was able to take all the sweat and confusion of the mine with a grain of salt. As he put it, the pay was the same whether you did three different things once or the same thing three different times. So why worry about it? As far as he and Luke were concerned, if a brass band marched down the drifts, I doubt whether they would have raised an eyebrow.

Day shift was positively the worst underground, not only because of the higher temperatures, but because the mine would be filled with all sorts of people who had little to do but get in the way and get on everyone's nerves. These included geologists crawling around with rock picks, engineers and surveyors checking angles and shooting grades to align drifts, the steady day's electrical and mechanical crews, and, of course, the shiny white hard hats of upper management. On days like that my shift boss would plant new ulcers and age visibly.

The ST-8s roared around the drifts and up and down the declines and day by day the decline headings crept farther down into the solid rock. Water accumulation was a problem at the face, and pumps were necessary to keep the headings free of water. Steel sets were used in place of timber because of the size of the drifts. The ST-8s hauled the big, clumsy beams into the declines, then with their big buckets lifted the men who would bolt the beams together and finish the timbering. For all their power and flexibility, the ST-8s were not at all infallible. It was fairly easy to get one of the big loaders stuck on the downhill grade while mucking out a heading after a shot. Once the front wheels were bogged down, getting the machine out could be a major project. Every different operator had his own tricks for this, such as raising the bucket to shift weight, shoving timber under the wheels, and using timber stulls to brace the loader on the face or ribs.

Every once in a while, an ST-8 would really get itself stuck, and none of the usual techniques for freeing it would work. In that case, a second loader would be called for and some cable would be strung between the two. On one memorable occasion in the South decline, even the second loader didn't do the trick, and a third was sent for. With three operators and three or four men on the ground, the whole effort turned into a circus, although a very dangerous one. The men on the ground were probably exposed to the most danger, that of being crushed by one of the snaking, writhing ST-8s. The operators had their own problems, including watching for a broken cable. When the one-inch cables snapped under the tremendous load, they would lash back like a big rubber band. They were capable of taking a man's head off as if it weren't even there. The heat buildup from three roaring V-10 diesel engines was pretty bad, as was the concentration of exhaust emissions in the drift. The amount of dirt and smoke being poured into the air every second made cap lamps only twenty feet away seem a dull red glow. There are catalytic scrubbers on the exhaust systems of the ST-8s, of course, and management swore that those devices put nothing but "clean" emissions into the air. On good days, one could look up the decline and see the portal as a tiny red orb like the planet Mars. That red spot was actually the blue Arizona sky, but there was enough dirt, dust, and smoke in the decline air to color it red. On this particular day, when three ST-8s labored in unison, the red orb was further reduced to a dull brown, and the air seemed thick enough to cut. For an hour, I sat on the middle loader, surrounded by roaring diesel engines and stretched cables, just waiting patiently for something to happen. Finally, at the end of the shift, just when a record fourth loader was on its way, the three-piece, ninety-ton parade of cables, steel, and diesel fuel lurched and lumbered up the decline.

After shifts like that, it was a particular relief to ride the skip up the decline, to feel the flow of air against your face, to pass the portal and burst into the sunshine and clear, blue sky, and to fill your lungs with air, real air, as the skip eased to a stop on the ramp. Only those who spend eight hours a day in the confines of an underground mine realize fully just how expansive and broad the sky is. One of the simple joys of life was being able to take off those sweat-soaked clothes, those damned hot rubber boots with another inch of sweat slopping around the soles, muck out your nose, shower, put on some decent clothes, and head for the parking lot.

The speed limit had not yet been reduced nationally to fifty-five, but was legally sixty-five across the expanse of the Papago Reservation. Not that it made one bit of difference. After a shift when a crew filed out of the dry, it was only a matter of minutes until a line of cars was flying through the desert heading back for civilization. Since the Papago Reservation is certainly one of the least populated and least developed sections of the country, there were few turnoffs and little traffic to worry about. The slowest cars drove at seventy or seventy-five, and the guys who were really in a rush to get back to town were limited only to what their engines would do.

This was the Sonoran Desert at its best, mostly flat country broken with diminutive lines of ridges and mountains, and all uniformly covered with a sparse blanket of saguaro, cholla, mesquite, and ocotillo. Arizona Route 64 connects Casa Grande and the mine and makes as few turns as possible to complete its job. The only signs that told that outposts of civilization did indeed exist in this bone-dry desert were hand-painted boards marking dirt turnoffs leading to the tiny Papago settlements of Gu Komelik and Kohatk. On the longer straights, the strip of black asphalt would lose itself in shimmering silver mirages, only to appear once again before fading into the distant blue-purple ridges.

It amazed me that cattle could graze in that stuff, and there would always be a few white-faced Herefords moving through the saguaros. Since there wasn't a drop of free water anywhere, those cattle must have kept damn close tabs on their watering troughs. I imagined those white faces would have gone crazy if they were suddenly herded into a lush Virginia meadow with six-inch-high green grass and a bubbling brook flowing through the middle. Doubtlessly there were no tears shed in the government land offices when that enormous tract of desert land was set aside for the Papagos. If there had been known deposits of high-grade copper ore on that land, it is not likely the Papagos would have gotten it in the first place. But time works in strange ways, and now the high-grade deposits of Jerome and Bisbee are exhausted, so the mining industry has turned its attention to those lower-grade deposits that were once deemed uneconomical to work. Lakeshore is one of these, and the leases and royalties that will go to the Papagos will be handsome indeed. That all land is worth something, no matter how superficially valueless in appearance, certainly proved true here.

After eight hours of the underground heat, the crews had worked up monumental thirsts, even though they drank water constantly during the shift. On the hottest days, depending on one's exertion, a miner could drink a quart of water every hour, or two gallons of water per shift, and not urinate once during the eight hours. All that water served only to maintain the fluid levels of the body which were rapidly being drained by heavy perspiring. When the shift was over, it wasn't water that many of the miners had in mind. The more innovative ones would have insulated ice chests loaded with beer waiting in the trunk, and the tops popped the instant that the mine property was cleared. The steady stream of expended aluminum cans that flew from the car windows caused someone to remark that driving behind a particularly hard-drinking carload was like watching the expended brass shell casings being ejected from an automatic cannon. All that aluminum stacked up off the shoulder of the road in a narrow band about six feet wide for many months, and the flagrant littering was rationalized as "enriching the high-grade aluminum deposit."

Many of the miners coming off shift would wind up in McNatt's Wonder Bar, a popular drinking establishment on Casa Grande's main street. The guesses as to how the Wonder Bar got its name were many.

Day shift has let out, and in a few minutes the bar will be three deep in miners. (ASARCO Incorporated)

Some said it was a wonder the air conditioning could keep up with the outside heat. Others said it was a wonder you could walk out of the place on a Saturday night with your teeth still in your head. The Wonder Bar had all the basic requirements: a loud, blaring juke box loaded exclusively with country and western, a long bar and quite a few tables to handle the crowd, a pool table, and an inexhaustible supply of tap beer. Cowboy prints, both serious and comical, lined the walls, and a six-foot spread of longhorns hovered over the cash register. Dangling on a string from one point of the horns was a cardboard placard listing the previous year's scores of the Arizona State football team. The Wonder Bar didn't take much getting used to; it offered no pretenses. It was understood that when the miners pushed open the door about forty-five minutes after the shift let out, they were there to drink and make noise. Every other song that blared out of that juke box was Jeannie Pruitt singing about her satin sheets, and the conditioned response I developed over that song remains to this day. Whenever I hear it, I immediately imagine walking across the burning parking lot in the glare of the afternoon sun, pushing open the door of the Wonder Bar and feeling the rush of cool air on my face. I remember pausing for a second to allow my eyes to accustom themselves to the gloom while that wailing soprano voice wore out the satin sheets, and then hearing another voice shout over the din of the conversation, "Hey Pard, pull up a chair, you been sittin' on your ass all shift anyway." It was also a wonder how just walking in that door could make the heat of the declines and the demonic roar of the ST-8s seem so far away.

There is little about Casa Grande that would distinguish it from any of a hundred towns in the West of the same size. With a lot of desert to grow into, the town tends to sprawl rather than develop around a given center. I doubt whether there is a building higher than two floors. In 1973, Casa Grande was in the middle of a boom, with two mines opening up and others who favored the location moving in. The location provided quick access to both Phoenix and Tucson, without making one put up with city life.

When I hired on, housing was almost non-existent, and every motel in town was full, many rooms taken up by the development crews of the contractors working for Hecla. For lack of anyplace else, I took a room in the San Carlos Hotel, a large Spanish-style stucco building that had seen better days fifty years ago. The hotel was located adjacent to the main line of the Southern Pacific Railroad, and, since the building lacked air conditioning, it was necessary to keep the windows open all night hoping for a stray breeze. Sleeping was next to impossible, as the Southern Pacific runs a lot of long freights between Los Angeles and

A drill jumbo requires only a few minutes to drill a hole ten feet deep and 1½ inches in diameter in solid rock. (ASARCO Incorporated)

El Paso. I think half of them were routed through the lobby of the hotel. There were a lot of Hecla hands at the hotel, and all of us spent a lot of off-time looking for better accommodations. Within two weeks I was fortunate to be at the right place at the right time and found a decent place, a change that made the mine shifts a little more bearable.

By midsummer it was hotter than ever, and, somehow, Hecla was getting more work than ever out of the decline crews. The declines were nearing sea level and were nearly 7,000 feet from the portals and 1,800 feet beneath the surface of the desert. In an idea to get still more work out of the miners, management decided to experiment with a contract in the decline headings. This was naturally met with approval from the miners, who reasoned that, if they were going to break their backs anyway, they may as well have a shot at some more money. One of the decline miners, Mark Lorentine, thought so much of the idea he became a little overly enthusiastic.

Mark was a full-blooded Papago, and one of the best miners Hecla had, having worked on the declines from their start. At the time the contract was announced, only six-foot rounds were being drilled and blasted, although the diesel-powered jumbos were capable of taking out a ten-foot round. The shorter rounds had been ordered by management, since the rock wasn't the best and the miners would have to contend with less unprotected back. On the first day the contract went into effect, Mark thought it over and decided that it would sure be a waste of good contract time, and therefore money, to stop those ten-foot drill steels at six feet. It was just too much of an opportunity to pass up. Mark drilled out a ten-foot round, loaded it and shot it. He figured that nobody would be any the wiser, and that it sure would look good on the footage report. When the smoke had cleared, unfortunately, it looked like a banquet hall down there. Management was upset, time was lost, and the contract, possibly the shortest on record, was terminated.

Mark had no idea how that could have happened. "Just the rock, I guess," he told them as stoicly as ever. But when the white hats turned to leave, Mark allowed himself a smile and imparted a bit of Papago philosophy. "See," he said in a very matter-of-fact tone, as though he had just confirmed a basic truth, "you'd always be wondering whether or not that ground could take a ten foot round. Now you know."

Maintenance on the ST-8s seemed only to become more of a problem, and it was unusual for a loader to last through eight solid hours of running without something going wrong with it. Hydraulic leaks were a common problem, and often hundreds of gallons of the red fluid would leak all over the drifts. Seals and hoses would go at any time, and an operator would suddenly be confronted by a fountain of hot oil spraying six feet in the air from some part of his loader. Too low a level of hydraulic fluid on the declines could mean serious trouble, even the loss of gears, which, although not a common occurrence, did happen. The operators were always able to steer the loaders into the rib to stop them. Then, it was simply a matter of getting to a phone and, if these were out of order as usual, making the long, tiresome walk up to 500 and reporting the trouble to the mechanics.

"Where is it this time?" the mechanic would ask.

"Down the South decline, below 350."

"What's wrong with it?"

"No hydraulic fluid, no gears, no nothin'."

He'd shake his head, wipe the grease from his hands, and ask in an accusing tone, "Did you check the level when you took it out of here this morning?"

"Yeah, it was up then. Guess it leaked out."

"What do you guys do down there? Drink the stuff?"

I could sympathize with the mechanics. The ST-8s took a beating in the course of a shift and the mechanics, between keeping up with the regular maintenance and repairing the constant breakdowns, had their hands full. Just about the time I'd sit down at a table to wait for the mechanic to round up a helper and load a jeep with tools and a drum of hydraulic fluid, my shift boss would walk into the bay. He'd take one look at me sitting there, stop short, and a look of suppressed outrage would cross his face.

"Why ain't you down the declines, Pard?"

"Loader broke down."

The new look on his face would be one of utter and complete frustration, as though the whole rotten world was against him in one concerted effort. "Jesus, Jesus, Jesus," he'd say in honest anguish. "What are you going to do? Just sit on your ass?"

What do you want me to do? A prayer dance around the loader? "Pard," I'd say in a soothing voice hoping he'd calm down, "as soon as the mechanic gets his gear together, we're all going back down and get it running."

Turning to the harassed mechanic, he'd ask, "Won't take long to get 'er runnin', will it?"

The mechanic would hesitate before answering, as if consciously formulating a civil answer, then say in measured words, "If I can fix it quick, then it won't take long. If I can't fix it quick, well, then I imagine it'll take a little longer. First, I guess I should find out what the problem is." Like, what do you think I am, a goddamn magician?

"Oh, I understand that alright, Pard. It's just that I got so damn many things to get done . . ."

"Really? That's funny, I haven't had a thing to do all week."

With the point made, my shift boss would stalk out of the bay to grow his ulcers elsewhere, leaving the mechanic to finish loading the jeep. I liked the way the mechanic looked at things. It would run when he could fix it, and that was that.

Sometimes an ST-8 could really get the adrenaline going. I was told to take one loaded with scrap steel from 500 to the surface on the South decline. This was a long pull, since the loaders would only do about two miles an hour in low gear and it was 5,600 feet, more than a mile, to the surface. As the mine air grew thinner up the declines, the air-cooled engines, which faced the portal, tended to run in their own pocket of hot air and overheat. If nothing were done, the engines would simply quit, a bad thing to have happen on the decline. Several crosscuts between the declines could be used as level areas for pulling the loaders off before continuing the run.

I was quite settled in the operator's seat, as comfortably as one can get, listening to the monotonous, yet reassuring, drone of the engine. The uphill progress is painfully slow, as the pipes and different rock shapes creep past and the distant sphere of light that was the surface began to get larger. I had pulled off twice to cool the engine down and was on the last leg to the portal when the temperature needles began climbing again. With less than one hundred yards to go before reaching level ground within the concrete portal, the red temperature warning lights blinked on. I was close enough to the portal to see the level, which, as a true horizon, shows just how steep the declines are. It was like looking at the crest of a steep hill, so near, so I decided to keep going rather than shut the engine down on the grade, all the time hoping the engine wouldn't shut itself down. Quite suddenly everything went to hell. Less than a hundred feet from the level, the engine erupted in a big orange ball of fire. Very unpleasant things flashed through my mind. I imagined the hundred or so gallons of fuel in the tanks and the glorious possibility of riding a ball of fire down the decline at a hundred miles per hour. I smashed the red fire button on the panel and a huge cloud of white fire retardant powder shot all over the engine. Between red balls of fire and white clouds of powder, I thought my hair was going to turn white. The fire system succeeded in killing the open flames but did little for the dense smoke that billowed out of the engine compartment. The last fifty feet seemed like fifty miles, but the loader finally reached the level. I took it just beyond the portal until I had the warm sun and deep blue sky over my head, then set the air brakes, cut the engine, and bailed off the damn thing. I was sure it was going to blow itself and me to hell at any given moment.

From a hundred feet away, I watched the last wisps of black smoke rise from the blackened engine compartment of the big yellow ST-8. The cloud of ominous smoke which had heralded my arrival and emergence from the portal had attracted quite a few curious onlookers from the surface crews. Again I had the chance to walk into the mechanic's shop, this time on the surface, and to inform them that another of their loaders needed tender care. A whole crew of mechanics walked out to the stricken loader, and, after a cursory examination, pronounced the cause of the fire to have been spontaneous combustion of the heavy layer of grease that had accumulated on the outside of the engine. No one seemed the least bit perturbed by the whole thing, and the mechanic shift boss told me that as soon as they could recharge the fire system, I was free to take 'er right back down the declines again. That was very disappointing news indeed, for I had been certain that such a near-catastrophe would put the offending loader in the shop for at least several shifts. But Hecla never wasted much time in such matters, and, without so much as steaming the accumulation of grease off the engine, the fire system was recharged. Half an hour later I found myself rolling the same smoke-blackened loader down the decline to get back to work.

The fact of the matter was that, in an underground mine, no one person could independently rely on his own skill, know-how, or experience to keep out of trouble. As far as the miner was concerned, there were three shifts, his and two others. All worked in the same headings or raises, and any error in judgment or deficiency in the workings they had constructed could bring the whole thing down on one's head. Men on all shifts would be placing their trust in what others had done. This mutual responsibility extended beyond the headings and the work of the miners; it included the mechanics, of course, since men relied on the loaders and other equipment which they had maintained and repaired.

The realm of responsibility extended even further than the obvious dependence between the miners and the mechanics, and into areas one might not think of. One such group upon which the welfare of all depended considerably was the engineers and surveyors. For some strange reason, many believe unquestioningly that the blueprints and calculations of the engineers, which are based upon the physical shooting of angles and measuring of distances by the surveying crews underground, are absolutely accurate. Perhaps that presupposition is based on the fact that these men are supposedly better educated, that they spend only a minimal amount of time underground, and that they are not paid to sweat and grovel in the muck. But these men are also subject to human fallibility. Although they sit in air conditioned comfort on the surface, their errors may very well affect the lives of the miners who are sweating and cursing in the heat and darkness underground and relying totally on their competence. An event that clearly demonstrated this occurred at Hecla in the summer of 1973. It acutely embarrassed both the engineering department and management. Very, very fortunately, by a matter of inches, it was destined only to be an embarrassment, and not the ultimate ignominy.

To understand the layout of a modern underground mine, one must visualize a three-dimensional model. There are the declines, progressing steadily downward at a constant angle, the vertical shafts, and the levels, running horizontal to their many branches. There are also various crosscuts, ore passes, manways, stopes, and ventilation drifts that may run uphill and downhill like a roller coaster while all the time turning left and right. It is only through applied mathematics, coupled with the techniques of surveying, that the precise relative locations of these workings may be projected on a flat piece of paper. To a layman, the complete layout of a modern mine will, in all probability, be merely a confusing, unrelated maze of workings. To mining engineers and surveyors, of course, the accurate plotting of existing workings and the accurate planning of future workings are expected to be well within their realm of competence, just as a miner is expected to know how to drill, load, and shoot.

Beginning at the 500 Level, a ventilation drift "took off" and wove its way beneath the progressing declines. A jeep ride in that drift was similar to one on a tight, winding mountain road, only underground. At the start of this particular shift, we were told that the previous crew had gotten started drilling a round in the face of the North decline. We were to finish the drilling, load it and shoot, muck out, and begin standing steel, following the pattern with which we were all so familiar. I was working with Dan Harger, a miner who had gotten to Hecla from Indiana by way of the copper mines at Superior, Arizona. As most everyone had some sort of a nickname, Dan's was Shag, in tribute to his bald head. But although the top of his head bore no hair, he made up for it with a profuse growth on his chin. His apparent disregard for the normal place upon which to grow hair became the butt of many jokes, and Shag was asked at least once a shift why his father hadn't taught him the proper place to grow hair. To Shag, all this was like water off a duck's back. He further enhanced his position as something out of the ordinary by living in Apache Junction and driving an evil-looking black motorcycle seventy miles through the desert to get to work. Shag was never late, and I doubt whether it ever took him an hour to get home.

When we got to the decline heading, we found the jumbo in position and the top half of the face drilled out. Shag took the left drill and I took the right, and in a few minutes the drift reverberated with the pounding roar of the twin rock drills. We had been taking ten-foot rounds out now for some time, so we both ran the drills forward for their full length. On the extreme bottom right holes, I noticed the drill steel would travel the last foot into the rock very quickly, as if there were no resistance at all. I pointed this out to Shag through hand signals, since voice conversation was drowned out by the roar of the drills. He watched the steel surge forward and acknowledged it with a nod and a shrug. At the time, I thought we might have drilled into a vug, or cavity, in the rock, and my hopes turned to the possibility of high grade, recalling the crystals that were often recovered from similar formations at Climax. I made a mental note to thoroughly examine the muck pile after we shot. In less than an hour, we had completed the pattern of fifty holes, each ten feet deep. We then backed the jumbo out of the heading and up the decline, loaded six cases of dynamite in the loader, and ran it to the face.

I don't think anyone ever questioned the fact that the decline faces were the deepest part of the mine at that stage in development. There were no other crews to worry about down there, just you and the rock. Shag and I made up the primer sticks, inserting the appropriate time delay detonator caps in the ends of the 1½-inch sticks of powder. We used a ten-foot wooden loading stick to tamp the primers firmly at the base of the holes, allowing the remaining length of the twin electric lead wires to dangle freely down the face. When I worked my way down to the lower right corner of the face, I shoved a primer into a hole and punched it forward with the stick. To my surprise, the stick slid entirely into the hole until the end was flush with the face. That, of course, assuming everything is normal is a physical impossibility, since a ten-foot loading stick with 1½ feet of powder in front of it should stop after only 8½ feet.

"Yeah, there's some kind of hole back in here, Shag," I said. The damn stick went in all the way."

Shag, imperturbable as ever, was not impressed in the least. He was sitting behind me making up the last of the primers. "Good."

I tried to pull the primer out of the hole by the lead wires but couldn't. "Got it caught back there," I said. "I'm telling you Shag, there's a big hole back there, that's right where the drill punched through. We're gonna find some highgrade back here when we shoot it, I guarantee it."

Dan Harger loading a flooded lifter
hole with an eighteen-inch stick
of powder. (Stephen M. Voynick)

"Good, my kids have been yellin' for something anyway," he said, still not really caring. Getting to his feet, he began tying the round in.

Shag already had the round wired in, and I was still playing with the stick of powder, trying to get it out of the hole. "Well, you gonna play with that thing all shift?" he asked, getting impatient to shoot. "Just shove some more powder in, she'll go."

He stood a few feet away and watched me with his "you could foul anything up" look. I was getting impatient myself, and had about decided to say the hell with it and finish loading the hole, wire it in, and get out. The two of us stood there alone in the heading, the dark and quiet broken only by our cap lamps' beams and the incessant drip-drip-drip of water. I had the lead wires pulled tight in my hand when I felt a small but distinct tug. Knowing there was nothing on the other side of that rock but China, I credited it to my imagination. Then I pulled a little more on the wires and, unmistakably, the wires pulled back, not with a steady withdrawing motion, but with a feeling that was very much alive, like a catfish taking a hand line. What the hell is this, I thought, knowing there had to be a good explanation. My thoughts ended abrutly when the lead wires gave three good, solid tugs. The hair stood up on the back of my neck.

"Hey," I said.

"Hey, what?" Shag asked in his impatient voice.

"Something pulled back on these leads."

He raised his eyebrows and rolled his eyes back a little. He thought I was nuts, and I couldn't blame him. "Good," he said with a big grin, willing to play my silly game. "What do you figure it is?"

"I'm not shittin' you, Shag, something pulled on this lead." My tone of voice told him that I was not playing a game, or else that I was a damn good actor. Just then I felt a strong, almost irresistible tug. My fingers were drawn nearly into the drill hole and I gave Shag what must have been a truly incredulous expression.

"Let me see that sonuvabitch," he said, now quite interested. He didn't have the lead wires in his hands two seconds when he looked up at me with wide eyes and jaw agape. "What the hell *is* that?"

I didn't know and I wasn't sure I wanted to find out. Anything that could chew through 1,800 feet of rock might not be anything I'd want to meet. Not down in that decline, anyway. I took my hard hat off, pressed my face against the collar of the hole and peered in. I saw some very faint light at the other end. "Turn off your light, Shag." With both our lights off, the heading was plunged into absolute darkness. I looked again into the mysterious drill hole and sure as hell, there was light on the other end. "There's light down there, Shag."

"Light?"

"Light."

"Lemme see," he said, pressing his eye to the drill hole. "Light," he confirmed a second later. We looked at each other in total confusion. "What the hell is goin' on here?"

"Anybody hear me?" I shouted into the hole, feeling more than a little foolish.

"Maybe you ought to talk in Chinese," Shag suggested, but his humor went for naught as a garbled voice resonated back through the drill hole.

"Where are you?" the distant, hollow voice inquired.

"North decline," I shouted back. "Where are you?"

If anybody had walked down to the heading at that moment and seen the two of us on our hands and knees, butts sticking up in the air, and talking to what appeared to be a drill hole, they would have thought we were utterly and completely crazy.

"Vent drift," echoed the answer.

"It's that vent drift, Shag. The vent drift! What the hell are they doing there?"

"Give those guys a jackleg and there's no tellin' what they'll do. Give 'em enough time and they'll come up in the cellar of the Wonder Bar," he joked. Then he became serious. "Pard, do you know what woulda happened if we shot this thing?"

I knew very well what would have happened. The detonation of six cases, three hundred pounds, of dynamite would have blown whomever was back there, wherever they were, to bits and pieces. "Well, I guess we ought to give somebody the good news. Boy, are they going to love this," I said.

They certainly did love it. The white hats, the engineers, and the surveyors crawled over the area and eventually figured out that the vent drift was much higher and farther east than they thought it was. The story had it that there were several people collecting their pay that afternoon. There wasn't much choice at that stage of the game but to shoot the decline heading. When the smoke had cleared and a little muck had been removed, there was an eight-foot-square passageway leading right down into the lost vent drift. It cost the company quite a bit of time and expense, since the whole mess had to be poured with concrete and straightened out. It was by only a matter of inches that two or three men weren't killed. If that vent drift had been a foot farther forward, or if a shorter round had been drilled, or any number of ifs, those last drill holes would not have broken through into the vent drift, and there would have been a mere foot of rock between the men in that drift and six cases of detonating dynamite. Either Shag or myself would have had the distinction of being the person to press the red button on the blasting box, and even though it technically wouldn't have been our fault, it would have been a hard thing to live with.

Miners will tend to accept an accident or a death much more readily if it has occurred simply, free of the stigma of incompetence or of gross error, in keeping with the nature of the hazards they accept and face every day. The Shelton accident, for example, was unquestioned and readily accepted since it involved the simple and common phenomenon of falling rock, an event that could happen at any given moment. Had the "Chinese drift" incident turned into a tragedy, there would have been a total loss of confidence and respect in management, and doubtlessly a lot of tramping to get out of what would have been branded forever as a "bad hole," since the accident would have been the product of inexcusable professional incompetence. One might soften the terminology by choosing "human error" as the explanation but, if that round had been fired, the result would have been exactly the same.

The miners at Hecla didn't enjoy the comfort of a lunchroom, since there wasn't any. Miners would simply gather at various points and make themselves comfortable on a stack of timber or a pile of muck, and the lunch ritual would begin. In the oppressive heat of the drifts, conversation would be carried on using only the shortest of staccato sentences, as if it were too much of an effort to waste words. Hard hats and lamps would rest on the ground by each miner's side, eerily illuminating each sweaty, grimy face. Each different voice would sound monotonously similar as the tone and individual qualities would become homogenized in the hollow resonance of the drifts.

"You guys find any more Chinese drifts down there?"

"Nah, nothin' today. There were some little yellow, slant eyed fellas runnin' around down there this mornin', though."

"They say what they wanted?"

"Lookin' for work. They heard this was one of the sharpest mines in Arizona and they wanted to get some of the action."

"How did you know what they was sayin' if they was Chinese? You talk their language?"

"No dummy, I talk to 'em in Okie. Everybody understands Okie."

"I'm surprised they didn't go back where they come from when they seen you was an Okie."

"Chinese are smart. They know good people when they see 'em." That would be followed by a chorus of low moans.

"You know, thinkin' about the idea of puttin' them Chinese to work down there, it just might be a good idea," someone else would add. "They just might have a good eye for that decline work, since they're both slanted." Another chorus of low moans and boos.

"Even from you that was a low-grade joke."

"Speakin' of low-grade, that air down here this morning is pretty low-grade. Worse than usual."

"Luke says they're puttin' that booster fan in pretty soon."

"Hell, they been talkin' about that fan for a month and it still ain't in."

"For what little work you do, I don't see what you need air for anyway."

"Twice what you do."

"If you did twice what I did, this goddamn place would be in production by now. And the rest of these clowns would be laid off."

"They work like they're laid off now."

"Hey, who wants some hot peppers?" That was Gilbert Verdugo, one of the Mexican-Americans, a group that was collectively classified as "Mexicans." Gilbert's reputation was based not on his mining, but rather on the enormous Mexican lunches he would pack.

"Gilbert, you wouldn't know a hot pepper if you tripped over a pile of 'em."

"No, these are good, c'mon, you ought to try one."

"Hot as it is down here, I don't know how you can eat that stuff."

"I'm tellin' you, these are really good. I get 'em from my grandmother. She grows 'em down around the border, by Sasabe."

"You're really a wetback, ain't you, Gilbert?"

"Better'n a Polack."

Another voice. "Both a little better'n Okies."

"I'm tellin' you guys, these peppers are good. Trouble with all of you is you ain't got no balls."

"Who ain't got no balls? Gimme one of them things."

"Put the whole thing in your mouth at once," Gilbert always suggested, as he extended the waxpaper with the pile of dead-looking green peppers on it.

"I know how to eat. I done it before." With a delicate motion that seemed so out of place in the underground, the anonymous miner would bite the pepper off the stem and flick the stem into the darkness.

"Well? You like 'em?"

"Well, Gilbert, I'll tell you," pausing for a few slow, unhurried chews, "I don't know why you call these things hot. Where I come from, we grind stuff like this up for baby food."

"Why you sweatin', then?"

"I'm sweatin' because it happens to be 110 in this goddamn drift. Why the hell do you think I'm sweatin'?"

"That why your eyes are tearin' too?"

"My eyes are tearin' because I cry every time I think of gettin' stuck with you for a pardner."

"Well, here, if you like 'em so much, take another one."

"No thanks," pushing his lunchpail back out of the way and lying back on a piece of lagging. "I'm full."

"Full of bullshit, maybe. You just can't handle a good Mexican pepper, that's all."

The chorus of chuckles would gradually die out as, one by one, the miners would assume the familiar leaning rest position with hard hats covering faces. The darkness, the heavy, humid air, and the rhythmical dripping of water would lull the crew off to sleep.

The silence would be broken by a splashing sound and a scuffle in the muck. "You sonuvabitch!"

"What'd he do?"

"Ahh, he dumped a bucket of ice water on my neck. Keep that crap up and you're goin' to be diggin' that bucket out of your ass."

"Just keepin' you alert, Pard. You ain't supposed to sleep down here. Regulations."

Silence again. The miners who are awake are thinking about the remaining 3½ hours of the shift. The allotted half hour of lunch is already over and this is company time now. It feels much better, more relaxing and rewarding, to be lying down on company time.

A voice in the dark. "You guys remember when that super and that foreman drove down the South decline with their lights out that time? Sneaky bastards thought they was gonna catch us sleepin' or screwin' off 'cause they couldn't see no lights down there. We was all up in the crosscut gettin' ready to shoot. We lit that goddamn face off and the concussion and smoke come up outta there like a tornado. All you could see was cap lamps goin' on all over, coughin' and hackin', and that jeep backin' up as fast as it could. Served 'em right, the bastards."

No response, only a couple of anonymous murmurs. A few more precious minutes of rest on company time.

"Hey! Lights." Something would be rolling down the decline still far off in the distance and darkness.

"Better get movin'. No tellin' who the hell it is. Already ten of."

"I can't make it, Pard, go on without me."

"C'mon, c'mon, it's probably your shifter."

"What's this my shifter crap? He's your shifter, too, you know."

"Yeah, but he likes you better."

Coming to work or going home? The light behind
the miner is his partner. (ASARCO Incorporated)

Slowly, the cap lamps would start moving as hard hats were picked up and gear strapped on. Everything in slow motion. Once a man stopped in that heat, it was hell getting going again. The miners would pair up and move off into the darkness, the last bits of conversation getting lost in the shuffling of boots.

"Hey, how was that pepper Gilbert gave you, anyway?"

"Like a blowtorch for crissake, I still can't swallow. You got any water left?"

The ventilation, or rather the lack of it, at the Lakeshore mine was the single greatest discomfort underground. Depending on where a man was, it was either feast or famine. Directly under a vent shaft, strong gusts of fresh air would howl down the drifts to purge the ST-8 exhaust emissions and the dynamite fumes and smoke that constantly poisoned the air underground. Sometimes it would be hard to stand in the wind blast, and a common detail for two men was to hose down the drifts to reduce the dust that was kicked up. The declines, unfortunately, received no such rush of fresh air, and for ventilation relied only on the air that could be forced down in the big yellow vent bags that faithfully followed each drift and decline. The vent bags were big clumsy canvas tubes, and great pains were taken to keep them far away from the movement of the ST-8s when they were installed. This was often impossible, since the drifts were already cluttered with air, water, and sump pipes, electric lines, high voltage power supplies, and cables. The operators were duly instructed to avoid contact with the vent bags as much as possible. They did a fairly good job, but some ripping of the bags was inevitable, and when an ST-8 snagged one of the bags, the rip was rarely less than six feet long, if the bag was still hanging at all. When a decline bag was ripped anywhere along its thousand-foot length, the air reaching the headings was reduced to a faint breeze, and work was simply impossible. Sewing the bag back together with blasting wire became the next project. There were numerous places in the mine where it was physically impossible to drive an ST-8 without touching the bags and, where they weren't ripped, they were nearly worn through from the constant friction and rubbing of the big loaders.

The company engineering and safety departments, as well as the Bureau of Mines, had numerical standards for almost everything, including, of course, ventilation. This was based on a minimum flow of air in cubic feet per minute for a drift or working place of a given size. Temperature, which was always high, was not a consideration. We complained often about the air volume and quality that we received in the declines, not just for the sake of general complaining, but because we were almost physically incapable of continuing the exertion of heading work. To attempt to continue working in such conditions of poor ventilation will almost certainly result in splitting headaches, and if a man pushes hard enough, possible nausea and collapse. The complaints, most of the time, resulted in some technician's venturing down into the declines to hold an air flowmeter in front of the vent bag opening and proclaim we were receiving more than our minimum and, therefore, everything was wonderful. Everything was always wonderful on paper, because the people who put it on the paper didn't have to work in those conditions.

In an effort to divert what air was available to the working headings, ventilation would sometimes be cut off entirely to certain dead-end drifts not normally used in the course of a shift, including certain storage areas. These dead-air drifts would sometimes be roped off with a warning sign proclaiming "Danger — Unventilated Drift." The temperature in those dead-air drifts would rise in less than eight hours to something like 120°. We asked the safety men who measured the dust levels, air flows, and temperatures just what the actual temperatures were in the drifts when we worked. Apparently management had instructed them not to make such data common knowledge, going on the assumption that what we didn't know wouldn't hurt us, for they would just smile and walk away. Sometimes we discussed jumping them and taking their own temperatures rectally.

Along with the enormous drinking water consumption of the underground crews, salt was also a vital requirement. Although the company was good enough to provide salt tablet dispensers at different points in the mine, heavily salted sunflower seeds were the most common source of the salt that was rapidly sweated out. Most of the miners carried a bag of the seeds with them, and I would marvel at the oral dexterity that was necessary to separate the shells from the seeds using only the tongue and the teeth. If only one or two of the seeds are worked on at a time, this is no great feat, but the miners would stuff a handful into their mouths, a ten- or twenty-minute supply, then carry on a conversation and never miss a word, all the time spitting out shells as if it were a normal physiological process. The best indication of the actual volume of these things that was consumed in the course of a shift was the piles of shells that would stack up here and there. Little piles of sunflower shells at different points in the drifts told of a previous conversation that had taken place there, but the best accumulations were made by the operators who could spend an entire shift sitting on an ST-8. At the start of many shifts, it would actually be necessary to clean the pile of shells off the instrument panels in order to see the gauges.

During the Fourth of July break, one of my friends, Ted Wiswell, from Leadville, flew to Phoenix. Ted, whose pipeman's bag I used to stuff full of rocks, came partly for a vacation and partly to see whether it might be worth tramping Climax for Hecla. I met him at Sky Harbor on the afternoon of a good hot one that must have been about 110°. Heading south on the Interstate for Casa Grande, we made the usual small talk, and Ted remarked that he had had to scrape frost off the windshield of his car that same morning in Leadville. Although I had pretty much gotten used to the Arizona heat, at least as much as I was going to, the thought of frost and the absurd desire to jump head-long into the icy headwaters of the Arkansas River were making me uncomfortable. It was nothing compared to what Ted was going through. The car didn't have air conditioning, so we were moving at seventy miles per hour with the windows and vent open. Ted was quiet for a while, then exploded with, "You got the goddamn heater on in this thing?"

I thought I'd crack up. "Pard, that's the fresh air vent," I laughed.

"Crissake," he muttered, unbuttoning his shirt. "What the hell do they do out at that mine you work at? Air-condition it?"

"Yeah, they air-condition it, all right. Just like this car's air-conditioned." Ted just shook his head.

In less than two months at Helca, I could see that the Climax Molybdenum Company was in all respects a country club compared to the Lakeshore mine. This included safety and general working conditions. The only thing Hecla improved on was winter driving to get to work. The underground heat was more than merely uncomfortable and fatiguing. It was hell on the skin, and by the end of July I had picked up a skin rash that covered my back, chest, and arms. I played around with home remedies for a week or two and wasn't getting anywhere, so I finally decided to visit the local clinic and stand in line with the lame, the infirm, and the pregnant. A doctor examined me while simultaneously filling out the forms to make sure he'd get his from the company medical insurance plan.

"What kind of work do you do?" he asked absentmindedly between pokes, prods, and pinches.

"Miner. Underground."

"Uhuh, we're going to call this contact dermatitis. It's a condition brought about by dirt and excessive sweating. I'm going to give you a prescription which should reduce this pretty well. Now you'll have to keep your skin clean and avoid sweating for a week or so."

"That's going to be pretty hard to do," I said. "I told you I work as a miner in a filthy hole where it's 110° on the cool days." The doctor gave me a blank uncomprehending look. "Is there any way I can get a couple of days off?"

"I can't give you an industrial, if that's what you mean. This is only a skin rash. Just work somewhere for a week where you'll keep cleaner and cooler." Like, what's so hard to understand about that?

"There is no such place underground," I explained. The doctor shrugged. I walked out and picked up the prescriptions, debating whether I should even mention the issue at the mine. I decided it couldn't hurt and just might do some good.

"Hey, Pard," I said to my shift boss with an amiable smile at the start of the next shift. "I saw a doctor this morning about this rash I picked up, and he said it would be a good idea if I'd, go a little easy on the sweat and dirt for a couple of days."

"Lemme see your slip."

"He didn't give me one. He said it's not a full-fledged industrial, but if I watch it for a few days I can probably get rid of it."

He gave me an incredulous look. Malingerer. Unmitigated gall. We glared at each other for a few seconds while I debated breaking the lamp battery I was holding over his thick head. My bank account wasn't up to that level yet, so I decided it would be unwise to indulge in that great pleasure. I turned and walked away, strapped my lamp on, kicking myself for being so foolish as to ask. I did the best I could with the rash.

During August, when the underground temperatures were at their worst, I saw an article in the *Phoenix Republic* about the plight of the New York City clerical and office workers. It seems the air conditioning system had failed in one of the municipal office buildings in the Big Apple, and the inside temperatures had soared to over eighty degrees. It was impossible to work in such horrendous conditions, they had said, and with the backing of their union, they had walked off the job until the city could remedy the situation.

One would think there would be a limit to the injustices upon which the world is built. Doubtlessly, the everyday conditions the Hecla underground crews put up with had altered my point of view, but what actually was it that gave these New York office workers the moral justification to walk off the job? Surely it couldn't have been the fact that a little sweat would wash off their underarm deodorant, thus making them socially offensive to each other. How much sweat can anyone work up sitting in front of a desk anyway? More likely, their justification was that the city was no longer providing what they had come to expect. Not what they needed, but only what they had come to expect. The whole idea called for another beer as I re-read the article.

"Hey, Larry," I called over the din of the music and conversation in the Wonder Bar, "come here and read this." I was really curious as to how a miner would react to this. Larry Lunsford, who had been cursed with the misfortune of working on the shotcrete crew, ambled over and glanced through the article.

"What do you think of that?" I asked when he had finished.

"So what?"

"What do you think Hecla would do if we all walked off the job because it was too hot down there?"

"You know what they'd do," he said with a perfectly straight face, sliding the article back across the bar. "They'd fire every one of us and hire a new crew tomorrow."

He was absolutely right. Underground mining was one of the few fields left that retained to a large extent many of the work ethics that built this country. The only ones who benefit from it today, however, are the mining companies who use these antiquated work ethics to their best advantage. The article only served to point out how divergent underground mining and modern society have become in this country. While the overwhelming trend in labor is to push for often absurd creature comforts on the job, miners seem contented with the most basic safety standards, to the great joy of the companies. "Comfort" conditions in underground mines are largely impractical, of course, and as a result non-existent. The elemental safety standards provided by the companies are also still on the ground floor. The mining industry has made great strides since the 1880s, and many assume that the lot of the individual miner has similarly improved. This is a gross misconception, for his lot as a laborer has probably even retrogressed in comparison to the strides made by labor in general.

Miners' attitudes, in this respect, have also remained static, primarily because of the isolation of the field from the contemporary trends of labor. There is no free exchange of people and ideas in the mines; those who try it once and "get out" recall it only as a hazy memory of something they once did to make a buck, and as something that is no longer real or relevant. Those who make a life of it frequently "can't see the woods for the trees." It is all they have known and, not having any effective contemporary comparison of conditions elsewhere in labor, they feel no need to expect or to ask for anything better. They accept it all as simply a part of the game. The companies, of course, are more than willing to play that game.

Hecla had no intention of installing any cooling units on those vent shafts that served the development stages of the mine. Any cooling units would be installed permanently when the mine went into production. Thus they neatly avoided the expense of a temporary system. They realized very well that the declines would be driven and the mine developed regardless of what the conditions were. This is not the attitude solely of Hecla, but of the mining industry in general.

The drive to the mine on graveyard shift was quite pleasant. The desert at night was far less hostile and threatening than when it burned beneath the daytime summer sun. The bizarre, stark shapes of the saguaros, the ocotillo, the mesquite, and the ironwood trees that sank deep root systems to tap the water far beneath the bone dry creek beds took on a softness, almost a delicateness, in the pale light. A myriad of stars would glitter with a startling brilliance, and just off the road the devilish red eyes of the roaming javelina would burn like hot coals in the glare of the headlights. The sign marking the turnoff to the mine would appear too quickly on nights like that, and it would be time again to strap on the gear and brass in. It was so easy to just drive right past the gate and keep going, cross the border, and be in Guaymas when the sun came up. There were no mines in Guaymas, only a lot of sky and blue water.

The cement ramp at the portals would hold the warmth of the day, and it was quite comfortable to lie on it and stare up at the starry sky while waiting for the inevitable descent in the skip. This was as comfortable as it ever got on the desert in August, just before graveyard when the temperatures would drop into the nineties. The conversations on the ramp would be unusually subdued and quiet, as though everyone were consciously trying not to disturb the stillness of the surrounding desert. Many of the miners preferred graveyard over the other shifts, knowing that the air would be a little better and that there would be no white hats running around. Even the hardcore old timers could let slip a bit of mellowness and sensitivity in those last relaxing moments before the skip was ready to go down.

A few weeks before, I had written an article about the Lakeshore mine which was just published in the *Arizona Magazine,* the Sunday supplement of the *Phoenix Republic.* The fact that I played around with magazine writing was not one of the things I discussed regularly with the guys on the crew, and when the article was published it came as a surprise to many of them. The article was a simple account of what it took to put in a single shift underground at Lakeshore, and most of the guys read it, since it was a topic of common interest. Most found some mild incongruity in the fact that the dirty-faced guy who worked alongside them in the hole could also write magazine articles. A few were positively astounded. One was a young Mexican, Pepe, who, on the Monday after publication, stared at me with something approaching awe as I sprawled out on the concrete ramp to enjoy the last bit of fresh air and peace before going down. Finally, he shuffled over and said, "You really wrote that story, huh?"

I confirmed that I had, knowing very well what was going through his mind.

After a moment of silence, he asked, "How much money did they give you for it?"

"Hundred and a half."

He nodded. "How long did it take you to write it?"

"I don't know, maybe a day."

He shook his head slowly and solemnly in acknowledgment, and stared down at the concrete in thought. It was perfectly obvious what he was thinking. If this gringo can write stories for a hundred and a half a day, he must have a goddamn screw loose to work in this miserable mine. Finally he asked the inevitable question I had been patiently awaiting. "Okay, so if you can make that kind of money in one day writing stories, tell me why you work here."

I smiled and explained quietly, "Because unfortunately I don't write one of those articles every day."

"Why don't you?"

Damned good question, I thought. Why the hell don't I? "It doesn't work that way," I answered, as much for my sake as his. "For one thing, after one of these shifts, I'm usually too tired to think. I only write one of those things when I get hit with a bolt of inspiration."

"How often do you get hit?"

"The hits are very irregular."

"Yeah, but if you didn't work here, you wouldn't be so tired and you could write all the time." He shook his head, still perplexed. "Man, I don't understand you. With your brains, you could probably get any kind of a job, a good job, that you wanted. Why here, for crissake?"

"Pepe, I'm going to tell you something," I said, "and I want you to listen. A lot of people, you included, think that any clown who writes some crap and gets his name printed under the title is something special. You know, lots of brains, white shirt, clean hands, fancy office. Well, don't believe that, Pepe, because it's not that way. They're no better than anybody who works in this hole. You know all those stories you listen to at lunch? Well, articles are the same thing, only the guy who wrote them knew something about grammar and how to peddle them. The only difference between a writer and one of our guys down there telling a story is that the writer spreads it around with a pen instead of his mouth."

I rested my case and leaned back on the warm concrete, pleased with both the humbleness and the honesty of my answer. Pepe, however, was visibly disappointed. I guess he had wanted to hear something else to bolster his preconceived ideas on the nobility of writers. He apparently still harbored some, for in the small talk we made later, he didn't quite treat me the way he always had. His smile wasn't the same, the warmth born of equality was gone. In his eyes, I was no longer just one of the guys anymore, although he made a half-hearted effort to treat me as if I were. It made me uncomfortable and embarrassed me. I liked it better the other way.

Even at moments like that, with no work being done and everyone lulled into a false sense of security, accidents could materialize seemingly right out of the star-filled sky. Swing shift had already come up and was in the dry showering and getting ready to head home. Our crew had been held up from the skip while a cherrypicker crane prepared to lift a rail carriage off the decline tracks. That was fine with everyone, and I was personally hoping it would take half the shift, as I was quite content to lie on the ramp. The crane came out slowly and attached a cable to the midsection of the carriage, which was nothing more than two sets of rail wheels connected by a single steel I-beam that was used to haul new rail sections into the mine. When all was ready, the crew sat up to watch the show. The simple operation instantly turned into just that when the crane operator, either because he had a mental lapse or because he was nervous in front of the graveyard audience, failed to maintain his center of balance. Slowly, ever so slowly, the left-side wheels of the five-ton crane lifted off the ground, first just a few inches, then a foot, then a little more. There wasn't a sound out of our crew, which, sitting in perfect safety, was totally captivated by the tilting of the crane. Then, with a smashing and crunching of metal and a great cloud of dust, the crane slammed over on its side. At the same time, the cable holding the rail carriage broke and the steel carriage, now back on the decline and free, accelerated like a scared rabbit and disappeared with a clattering roar into the darkness of the decline at a hundred miles an hour. Not a word was spoken as the smashing and banging came from farther and farther down the declined tunnel. Only when there was complete silence and the humiliated crane operator crawled unhurt from his battered crane did a loud round of applause and cheering go up into the night.

The bravos and encores came to an abrupt halt when my shift boss jumped to his feet and gave us his patented cold stare. If looks could kill, the graveyard crew would all have been dead. This meant at least another hour sprawled out on the ramp while a small crew rode the skip slowly into the decline to check out the damage and pick up the pieces. There had been no chance of injury, since at shift change the mine was empty. Damage was limited to the crane, the rail carriage, which was bent into a pretzel, and some pipes and cables that were torn loose from the decline. Everyone laughed and joked, but everyone was also aware of what could have happened if there had been a crew working in the decline or an ST-8 operating. No one talked about that.

There was a lot of overtime if you wanted it, but after five days of sweat and bad air, not everyone was ready for it. More often than not, the shift bosses would have to coax miners to put in a day or two overtime, although Saturday and Sunday were, without a doubt, the best days to work underground. Skeleton crews took care of certain jobs that would otherwise have gotten in the way of regular development work or that simply could not be put off, such as maintaining the pumps that controlled the water in the declines. If the pumps were not maintained in the declines, nearly eighty feet of water would accumulate in as little as two or three days to completely inundate the headings. This lake could be drained in about two days of steady pumping. The weekend work was called pump watch, and it was just that, watching the pumps. If you were so lucky as to not have any mechanical problems, it was easy money.

One weekend job called for changing all the overhead light bulbs that illuminated the South decline, all 7,000 feet of it. I was teamed up with an electrician whose specialized skills were apparently necessary to screw the bulbs in and out. A helper was also there to hand him the bulbs. I drove an ST-8 with the electrician and his helper and a couple of cases of bulbs in the bucket, which was always raised nearly to the back. I would stop the ST-8 under each bulb and they would perform their appointed tasks while I put all the weight I could manage on the brake pedal to hold the loader on the decline grade. The air compressor couldn't keep up with the brake demands, and I frequently had to pull off on crosscuts to build pressure back to a safe level. Meanwhile, the electrician and his buddy were just having a great time laughing and joking up in the bucket. I was worried to death about their safety in that bucket, knowing full well the limitations of the loaders. As non-operators, they had absolutely no idea of the danger they were in. If there were any trouble with that loader, those two would have had a ringside seat for the finale.

Considering all the effort and exertion that went into one of those shifts, and considering the physical appearance of the crew at the end of eight hours—caked with grease, sweat, and muck—one would think that all off time would be spent resting. But somehow just the thought of getting away from the mine could instill amazing new strength in a person. After some grueling Friday shifts, a bunch of us would pile in a car and head straight for Mexico, speed across the reservation, cross the border at Lukeville whether it was open or not, and, after crossing the worst stretch of desert in the entire southwest, arrive at the blue water of the Gulf of California. The endless expanse of blue sky and sea, the clean air, the cold Mexican beer, and a don't-give-a-damn attitude would put that mine and everything related to it in a different world. We would try to time the return trip to coincide with the start of the shift on Monday.

Three days in Mexico would have gone by as though they never happened. Suddenly we would be standing in the dry, exhausted from lack of sleep, strapping on lamps, wondering where it all went. I remember Blue saying, "Pard, it sure seems like we start more shifts than we quit."

Damned if it didn't.

CHAPTER IV

FOR A PAYCHECK ON FRIDAY

HECLA MINING ZEROS IN ON LAKESHORE
COPPER PROJECT'S COMPLETION DATE

Hecla Mining Co.'s Lakeshore project is progressing at a rapidly increasing tempo, according to project manager James H. Hunter. The operation to mine and process a huge copper deposit located some 30 miles south of Casa Grande, Ariz., on the Papago Indian Reservation, has been in the development stage for more than three years. Construction, still underway, is being accomplished by Hecla management and labor forces, aided by a professional construction management team supplied by Ralph M. Parsons Co.

To achieve development of Lakeshore, Hecla has made significant increases in its labor force. Currently, 750 employees are on the payroll, with more to be added. In addition to its recruiting effort, the company has instituted a number of new programs, including the areas of training, safety, and security . . .

. . . Since January, 1970, some 6 miles of underground workings have been excavated . . .

"This Month in Mining"
Engineering/Mining Journal
August, 1973

How simple and clean it looks on paper, almost as if some mechanized wonder, remotely controlled from an air-conditioned office on the surface, were methodically eating its way through the rock. None of the sweat, mud, heat, and bad air. Just a nice, clinical, concisely worded statement, a report of the steady progress through the rock. Although there was still much work to be done before the mine would achieve its celebrated production status, the pattern of drifts on the main levels was beginning to assume its final configuration. The declines, by the middle of August, were driven beyond 7,000 feet from the portals, and were only about four hundred feet short of their termination point at sea level, 1,915 feet beneath the desert. The deeper the declines went, the worse the air became, a trend that would continue until they reached a twelve-foot-wide ventilation shaft at their terminus.

A completed vent shaft, No. 1, served as the primary air intake for the mine. This twelve-foot shaft dropped 1,200 feet from the surface to the 500 Level, and air was forced into the mine by two ninety-six-inch fans mounted in parallel at the collar on the surface. The intake air at the 500 Level was then picked up by screaming forty-eight-inch axial-vane fans and delivered with the use of the vent bag system and booster fans to the various headings and workings of the mine.

In August, two new vent shafts were well on the way to completion. These would greatly increase the volume of air to the 500 Level. One of these was No. 6, which was reached by an eighty-foot-long crosscut off of the 500 North drift. The total length of 500 North was 2,200 feet from its start at a four-way underground crossroads to the dead-end rock face where development had been temporarily halted. About 1,500 feet into the 500 North, the drift formed a Y, with the left leg being the eighty-foot-long crosscut leading to the floor of the No. 6 Vent Shaft, and the right being the continuation of 500 North and the remaining seven hundred feet to the face.

No. 6 had been started with the drilling of a small-diameter pilot hole from the surface to the end of the crosscut. A twelve-foot-wide reamer bit was taken underground and attached to the drill column. The enormous bit was then pulled back up to the surface, in the process reaming the pilot hole into a twelve-foot shaft. The huge load of cuttings produced by the reaming was allowed to remain in the shaft to provide wall support, while permanent concreting operations were begun from the surface. As the concreting progressed downward, the rock cuttings were to be mucked out from the floor of the shaft using ST-8s.

There were many easier jobs for the operators than mucking No. 6. It involved a run of about 1,500 feet, over a quarter mile, from the muck bay at the entrance to 500 North to the shaft crosscut. The muck, when it ran right, would stack up in a pile from the bottom of the shaft in a forty-five-degree angle to the drift floor. Since 500 North was on the level, the operators could use second gear on the loaders, which allowed considerably more speed, and the ST-8s would roar around the Y at full throttle and slam the big bucket into the muck pile, sinking it nearly to the hydraulic pistons. Then, with a combination of throttle and bucket manipulation, they would scoop a heaping load of the fine red rock cuttings, reverse gears, and start the quarter mile run back down 500 North to the muck bay. 500 North turns, bends, twists, and dips at different points to make it interesting for the operators. The run was further complicated by an enormous forty-eight-inch yellow vent bag and the usual confusion of air pressure, water, and sump pipes. Ear protectors were a necessity, as the operator's head would pass within a few feet of a screaming booster fan, and clouds of dust would be raised each time the loader passed one of the many rips in the vent bag, each with its escaping blast of air. The revving V-10 diesel engine would also pour out its share of abomination into the air. On the loaded runs to the muck bay, the speed of the ST-8 was equal roughly to that of the drift air exhaust, conveniently insuring that the operators would be in a constant cloud of black, hot, diesel exhaust emissions. When the operator finished his run out of 500 North, he would display his skills by simultaneously manipulating the throttle, brakes, gears, and bucket, an act requiring considerable practice and experience as well as both hands and feet, to produce a smooth, efficient dumping motion with the forty-ton loaded machine. When that was finished, he would point the big steel bucket at 500 North and repeat the process. It was necessary to run the ST-8s more than one mile just to remove about sixteen cubic yards of rock cuttings. Eight hours of this was a long, exhausting shift. If there was anything worse than running the declines, it was mucking that damned No. 6.

The concreting operations from the surface had progressed about four hundred feet into the shaft, and the top of the column of rock cuttings was only a few feet below that point, meaning that the total height of loose muck in the twelve-foot-diameter shaft was about eight hundred feet. To appreciate the actual volume, one must try to visualize a pile of muck on the surface twelve feet high and eight hundred feet long. The consistency of the cuttings in the reamed shaft was such that, if things went right, more of the material would simply move down the shaft to replace the removed bucket loads, thus maintaining the forty-five-degree angle on the muck pile. Things rarely went right all of the time, however, and one of the problems encountered in the routine mucking of No. 6 was that the muck would occasionally "hang up." That is, the rock cuttings would not fall to replace those which the operators had removed. If the operators continued to muck during a hangup, they would tend to undercut the entire column, creating a potentially dangerous situation. Several techniques were commonly employed to get the muck moving again after a hangup, among them playing a stream of water from a hose against the base of the shaft, some fourteen feet higher than the drift floor. That often worked, but when it didn't, a small dynamite bomb would be placed near the hangup and detonated. That usually did the trick. The muck would fall to fill the end of the crosscut, and the operators could begin hauling it out to the bay. A few of the operators would use the tips of their buckets in the extended position to scrape the base of the hung muck column. Often this trick would bring down enough for a load, but it had the disadvantage of placing the operator in a dangerous position if the whole thing were to let go at once.

I had the misfortune to be mucking No. 6 one hot August day, and I had undercut the column beneath a hangup. Since I was working alone, I didn't have the slightest intention of hosing the column or shoving some dynamite up there. If the column didn't want to come down according to the rules, well, the hell with it. It could stay there. I decided to get out what muck remained on the floor of the crosscut, a move which placed the loader half under the base of the shaft. That is exactly where it was when the muck column, without any warning, let go. All I could see in the illumination of the ST-8 headlights was a solid rushing wall of red muck, the weight and momentum of which was so great that it forced the bucket down, lifting the rear end, the engine end, of the twenty-eight-ton machine about six feet off the floor. In a second it was over. The ST-8 was covered almost back to where I sat with the moist, finely cut red muck. I managed to get the loader out, somehow, and it was only then that I realized how frightened I was. I decided then and there that it was time for an extended break. Muck falling like that was a new one on me, and I was quite impressed. When some jeep lights came down 500 North, and a shift boss got out to see what I was doing, I recounted the episode for him, thinking he too would be impressed.

"Yeah," he said with a casual smile, realizing this was the first time I had witnessed the thrilling event, "that'll happen now and then. You got to watch 'er, Pard. Don't get yourself too close to that sonuvabitch 'cause once she starts comin', there's no tellin' when she'll stop."

'Cause there's no tellin' when she'll stop. In just one week those words were to prove grimly prophetic.

It was incidents like these that would be brought up at the regular safety meetings. These meetings were conducted by the shift boss, and attendance of the entire crew was required. They were informal exchanges that were conducted wherever a shifter could corral his whole crew, sometimes no easy feat.

"Okay, okay, let's quiet down, we're gonna have us a safety meeting. Everybody's here, right?"

Moans and shuffling as everyone gets comfortable. Everyone complains and curses, but the truth of the matter is that this is on company time and there isn't a man on the crew who wouldn't rather lie around and feign interest than work.

"That's everybody, ain't it?" The shift boss looks around counting heads. "C'mon, straighten up, you ain't gonna sleep through this one."

"I think two guys are missing."

"Two? Who are they?"

"Those Chinese guys we been workin' with down there."

Snickers and guffaws. "Yeah, they're sneaky little bastards."

"Okay, okay, let's cut the crap and get serious." After checking a sweaty scrap of paper with his notes, the shift boss went on. "Okay, first. Now, you operators. When you pick a loader up at the start of a shift, you guys have got to check the levels. You drive these things all over the mine and then have to leave 'em somewhere cause there's no oil, no fuel, no hydraulic, no nothin' in 'em. You drive 'em down here on these declines and you know what that means. A little trouble can turn into big trouble. Especially that hydraulic, you lose them gears, you're gonna get screwed up. So check 'em good before you start, it don't take long."

"When are we gonna get some goddamn air down here?"

"Now just a minute, first things first. Now, a man on swing got hurt down here the first of the week, most of you know that. He fell off a jumbo and hurt his head and his back, had to be hauled out. Now all you know them jumbos are full of oil and grease, so just take the time to watch where you step. Just takes one step in a bad place on that slippery crap and you're gonna get your own ass hauled out."

"They oughta steam that equipment once in a while."

"Yeah, why the hell don't the mechanics or somebody clean that stuff off?"

"Okay, okay, I'll put it down for a suggestion." The shift boss fumbles around with a two inch pencil stub, scribbling on a corner of the greasy little worn-out piece of paper.

"What about the goddamn air?"

"Now just a minute, Pard, we'll get around to the air. I got some other things we're gonna go over first. Again, you operators, the electricians are doin' a lot of work up on that vent drift and they got high voltage cables layin' all over the place. Be careful when you take your loaders past there, cause if you tear one of them damn things up, you're gonna cook yourself." A pause to look around to see whether anybody is sleeping. "You sleepin', Nick? Nick!"

"Huh? Oh no, just restin' my eyes, Pard."

"Well, rest 'em off shift and pay attention to this. Might keep you from gettin' hurt. Now barrin' down. You guys still ain't takin' the time to bar these headings down right. You know this rock. You all know she's slabbin' a little. Crissake, take five minutes and bar it down right, it's your own goddamn heads."

"Half the time there ain't no goddamn bars down here."

"Yeah, what the hell do them other shifts do with all them bars, anyway?"

"Now, wait a minute, don't go blamin' everything on the other shifts. You guys are still throwin' your tools all over down there. You keep them headings a little neater and you'll be a little safer."

"What about the bars?"

"Okay, I'll see if we can get a couple more down here. But if we do, for crissake put 'em where you can find 'em."

"Now what about the goddamn air? Are we gonna get some air down in them declines or not?"

"Okay, the air. I know the air ain't all the time what it should be—"

"Ain't what it should be? There ain't enough air down there some days to light a damn cigarette. The deeper we get the worse it gets." A chorus of confirming grunts is the first indication that there is any group interest.

"I know it ain't good, but they're talkin' 'bout puttin' that booster fan in next week."

"It's always next week."

"It's been last week for the last two months. I'd like to breathe this week."

"Now take it easy. It's been a while, but this time they're gonna put it in 'cause they already got the fan down on 500. So they just gotta put 'er in. You know, you guys would all have a little more air if you operators would go a little easier on them vent bags. You tear 'em up faster than we can fix 'em."

"That's the asses on the other shifts."

"Yeah, why don't they hang them damn bags out of the way a little more? They take up half the damn drifts."

"The drifts are tight, Pard. If we all take it a little slower with the loaders, well, maybe we won't tear 'em up as much and we'll all have a little more air. Now, I got one more thing I'm gonna cover again and that's horseplay. I'm gonna tell you guys again, you have got to ease off on that horseplay before somebody gets hurt." The mention of horseplay always bring grins and muffled laughs. The shift boss pauses to indulge in a look of exasperation, then goes on. "Now it's not funny, not one damn bit. You'll think it's funny till somebody gets hurt. These cable ties you guys steal from the electricians, every time I turn around one of you has got somebody tied to the skip or a loader or some damn thing. One of you is gonna trip and fall and knock your goddamn teeth out and then it's gonna stop bein' so funny. So let's quit the screwin' around. Now, anybody got anything?"

"No. 6."

"Yeah. You operators, if you're haulin' muck outa that No. 6 borehole back on 500 North, don't get too far under that hole. When that muck let go, it can get dangerous, 'specially since you're back there alone. You get hurt, it's going to be a while before anybody knows it. So just muck what you can and if she hangs up we'll get a crew to get 'er goin' again."

"What did they do to those clowns that laid out the Chinese drift?"

"That ain't none of our business, Pard."

"The hell it ain't."

"Oh, yeah, one more thing. When you come in on Monday every one of you is goin' to be issued a self-rescuer. Now this is goin' to be your own personal piece of equipment, and you'll be responsible for it. It'll have your number stamped on it and you'll keep it in your basket when you go off shift. Now, they tell me that each one of these things costs forty-five dollars, so if you lose it, the forty-five dollars is goin' to come out of your paycheck. And whenever you're underground, you got to wear it. That's a law now. U.S. Bureau of Mines. So, on Monday, make sure you get issued a self-rescuer before you go underground."

The talk stops and it is apparent that everything of any importance has been said. Everybody is still sprawled out and in no mood to run back down the declines. The shift boss decides to be a nice guy for a change and not hustle the crew off to work immediately. A few more minutes of peace and quiet.

Finally the word is spoken. "If that's it, let's get back to work."

Silence.

"C'mon, this is company time. If we ain't talkin' about safety, we gotta get back to work."

Slowly the crew climbs to its feet, stretches, checks watches, and straps on the gear. Miners pair up and move off into the dark drifts. Suddenly there is a loud grunt and the clatter of a hard hat falling on the muck.

"What the . . . you sonuvabitch!" Gilbert is crawling around on his hands and knees. Someone had tied his lamp cord to an air pipe. With an electrician's cable tie, no less. The shift boss only stares at everyone on the crew with a disgusted look. He is looking for a telltale grin. The crew looks back with blank, bored expressions.

As promised, at the start of day shift on Monday morning every man working underground was issued a self-rescuer. The sealed canisters, about the size of a one-pound coffee can, were designed to be worn on the safety belt, alongside the already heavy tools and lamp battery everyone carried. Their purpose was to provide each miner

Mark Lorentine taking a water break in the heat of the underground. (Stephen M. Voynick)

A miner wearing a self-rescuer. These devices helped save 110 men during the entrapment and fire at Hecla's Lakeshore Project. (ASARCO Incorporated)

with some short-term protection against lethal accumulations of carbon monoxide from an underground fire. This was accomplished through the use of a filter containing a catalyst which could convert the deadly carbon monoxide to harmless carbon dioxide, a normal product of human respiration. The only drawback, unfortunately, was that the process generated considerable heat, just how much depending on the actual concentrations of carbon monoxide in the air. The self-rescuer would not remain effective in heavy concentrations of carbon monoxide for more than an hour. It was not intended to be a permanent life-sustaining device, merely a card in the hole which a miner might use to buy the precious time needed to evacuate a mine or seek refuge in a safe area.

The issuing of these self-rescuers, and the law making them mandatory for every person underground, were direct outgrowths of the Sunshine mine disaster on May 2, 1972, only one year and three months earlier. The day shift of the Sunshine comprised 173 men who showed up at their dry as usual to change into their gear, just as we were doing on the morning we received the self-rescuers. They spent the first half of their shift without incident, then paused for their usual lunch break at 11:00 a.m. As they resumed their work with half of the shift over, doubtlessly their thoughts turned to the other world, the bright world above to which they would be returning in a few short hours, and to the pleasant thoughts of taking the wife shopping, maybe a little fishing before it got dark, or having a couple cold beers. Exactly what happened then is still not known and probably never will be known. In a matter of a few minutes, the pleasant name of the Sunshine had become a cruel misnomer, for many of her miners were never again to feel the warmth or see the brightness of the sun in the spring sky overhead. Fire, probably caused by spontaneous combustion in old timbers, had suddenly released clouds of dense smoke accompanied by high concentrations of carbon monoxide that were borne along quickly on the mine's ventilation system. As underground personnel searched the workings in vain for the source of the smoke, the fire had only intensified, and the order was given to evacuate the mine. When the first cage-load of men had been lifted to one of the upper interior levels which eventually led to the surface, the hoistman, the sole man who could operate the cage, was overcome by carbon monoxide and died. Most of the miners on the lower levels, at the first sign of the smoke, had moved to the shaft station to wait for the cage. But a dead man cannot operate a hoist, and all the others had died waiting for the cage that never came.

The self-rescuers, basically the same device that we were being issued, were also present in the Sunshine on the day of the disaster. In

1972, however, mine safety regulations did not require the devices to be supplied by the company, and the Sunshine happened to be the only mine in the entire Coeur d'Alene district which had them available. The self-rescuers were not carried by each Sunshine miner, but were stocked at various points around the mine, just as they had been at Hecla. Later testimony showed that many miners had difficulty making the devices operable, and many others were forced to take them out of their mouths because of the extreme temperature of the air they had to breathe. A representative of the manufacturer of these self-rescuers also testified that the devices do indeed become hot, so hot that it causes a feeling of not getting enough oxygen and a tendency to cough. A person unfamiliar with these normal functions would quite possibly think he was inhaling smoke and would discard the device, as a number of Sunshine miners did. The representative stated that if the concentration of carbon monoxide in the drift air reached two percent, the temperatures of the air in the self rescuers could easily reach 450 degrees F. "Yes, it's hot," he said, "but the alternative is quick death."

When the full extent of the Sunshine tragedy had been determined, the death toll stood at ninety-one men out of the shift total of 173. It was a disaster second only in U. S. hardrock mining history to the copper mine fire in Butte, Montana, in 1917, which killed 163 men. The rescue operation at Sunshine was recognized as the largest in the history of U. S. mining and, in the end, proved worth it all, as two miners were brought to the surface alive after eight long days of entrapment.

An article about the Sunshine disaster in the *New York Times* on May 7, 1972, carried the following line:

> . . . *The attitude of mine officials was expressed by Bob Launhardt, the chief safety engineer: "If someone had told me, prior to this happening, that we would have a fire that would engulf a major portion of our mine in a matter of minutes, I would have said it was impossible."*

Bad underground accidents are always impossible. Right up until the time they happen.

The regulations introduced as a result of the Sunshine disaster included precautions to maintain a fresh-air environment, emergency bottled air supplies in the hoistrooms, the mandatory individual self-rescuer requirement, investigations to seek more suitable units, and mandatory training in their use for all miners. And so it was that on August 13, 1973, every underground worker at the Hecla Lakeshore project was issued a personal self-rescuer and trained in its use. The units had a simple mechanical seal that, once broken, would allow the top and bottom sections of the canister to fall away, leaving only a mouthpiece, similar to a scuba mouthpiece and a white cloth sack containing the conversion chemicals. A nose clip was also provided to prevent the inadvertant inhalation of carbon monoxide and smoke through the nose.

The crew fumbled around with their already crowded belts, trying to find the best place to attach the self-rescuers. The ST-8 operators particularly didn't like the devices, since there was little room in the loader seats, and the tools and lamp batteries caused enough discomfort.

"Where the hell you gonna wear this goddamn thing?"

"Ahh, I don't know, it don't fit nowhere."

"Make you a suggestion, Pard, Why don't you tie it around your neck?"

More fumbling and cursing. Ten miners standing around playing with their belts while the shift boss stands impatiently at the dry door leading out to the portal. Everyone is mildly upset at having to change around belts, but there is still time for the wisecracks.

"I'm tellin' you, Pard, tie the sonuvabitch around your neck. Your trouble is that you're so goddamn skinny, you ain't got the room on your belt like us muscular fellas got."

"I got a suggestion for you too, Pard. You can take yours and stick it up your fat ass."

By Friday, most of the crew had gotten used to the clumsy bulge on the hip opposite the lamp battery, and the once-shiny metallic surfaces of the self-rescuers were now well-covered with grease and grime. To the operators, they were literally and figuratively a pain in the ass. The novelty of carrying the self-rescuers had about disappeared along with the jokes. The declines got deeper and the air got worse, everything was status quo. The only good thing about today was that it was Friday. The weekend coming up meant you could either put in some easy overtime money or forget that the place existed for two days.

The authorized lunch period started at 11:30 a.m. on day shift and lasted for a half hour, but naturally if some miners worked in an isolated work area such as the South decline, and no shift boss was running around, a half-hour lunch was unheard of. On this particular Friday, the 19th of August, we were going to make it a good one. About twenty of the hour, the three of us just sort of slowed down and sat in the rib. Pretty soon the lunchpails were out.

"Little early, ain't it?"

"Hell with 'em. If anybody comes down here now we'll just tell 'em we stood the steel and there ain't no sense in runnin' around tryin' to find timber before lunch. They ain't gonna say nothin'."

"Don't kid yourself."

"Relax, everybody's probably eating lunch up there anyway. You think those guys take only a half hour?"

We all got comfortable in the muck and there was silence in the heading. The only sounds were the clinking of coffee cups and the rustle of waxpaper against the muffled, rhythmical throbbing of an air pump.

"Want to shut that pump up?"

"Ain't botherin' nothin'."

"That hole up there is goin' to take a lot of cribbin', you know."

"Yeah, I know. Worry about it after lunch. We'll all go up with the loader and get what we need."

"Friday today."

"Damn good thing."

"You guys comin' in tomorrow?"

"Ought to, need all the bread I can get my hands on right now."

"You?"

"Nope. I get enough of this place in five days."

"Hey, will you look at that."

"Look at what?"

"Up the decline. Jeep comin'."

"Shit. Those guys can read minds, I swear it. They never show up when you're breakin' your back, take one second to sit on your ass and they're all over the place."

"You know it's only quarter to, don't you?"

"Yeah, he'll say something. Christ, I'm glad today's Friday."

The bouncing lights of the jeep slowly made their way down the rutted floor of the decline. I could tell from the outline of the driver who it was before it reached us. Leo Yates, the shift foreman, stopped the jeep in front of us but did not get out.

"Little early, ain't it?" he asked.

"A little. Figured we'd eat before we rustled the timber."

Yates remained impassive. He asked quietly, "You three men are all operators, right?"

The answer came as three restrained affirmative nods. This sounded like a volunteer job in the offing, and experience told one never to volunteer unless he knew exactly what he was volunteering for. Right now we didn't know.

"I need some operators up on 500." Very calm, almost as though he were tired. Nothing urgent in the voice.

The three of us exchanged glances, our faces blank but our eyes smiling knowingly. No volunteers. An awkward second or two. Then the silence was shattered with the roar of Yates' voice, furious and full of authority.

"Goddamn it! Now! I got two men trapped up there and I need all the operators I can get! Get off your fuckin' asses and get in this goddamn jeep, the three of you!"

Lunch was over. There was no question that something somewhere was seriously wrong. The three of us left the lunches lying in the drift and climbed into the jeep. Yates put it into reverse and started backing up the decline. No one said anything until we reached the first crosscut and Yates turned the jeep around for the long trip up to 500.

"What happened, Leo?"

"They had a muck run," Yates said, again very quietly. "A big one. No. 6 back on 500 North. Hazelhurst got out but his two partners didn't and got stuck back in the dead end. They were tryin' to get a hangup to move and it must've come all at once."

I thought back to what had happened to me at No. 6 just a week ago and got a queasy feeling in the pit of my stomach. "Who's trapped back there?" I asked.

"Deeder and Udall."

I had only met David Deeder, a relief foreman, once, and didn't know him well. Terry Udall—Blue—I knew very well. When the jeep arrived at the muck bay that marked the beginning of 500 North, thirty or forty miners had already been working for nearly an hour. There was pitifully little any of them could do but stand around and wait for the single ST-8 that was somewhere in 500 North to finish its run, bringing with it eight cubic yards of the muck that was trapping David Deeder and Terry Udall. It was nearly 11:00 a.m. when we arrived at 500 North, about an hour after the muck run had occurred. The ST-8s worked steadily during that hour, but with a half-mile run required to remove a single bucketfull, only about six or seven feet of muck had been removed. The long length of 500 North that led to the blockage could accommodate only one loader at a time, and progress had been sadly, but understandably, slow. In the early moments of the accident, there was quite a bit of confusion as to exactly how much muck was blocking 500 North, but I don't think there was a man on the level who didn't believe that it was only a matter of time until the two trapped miners were freed.

That shift of Friday, August 17, had started as usual for Hazelhurst, Deeder, and Udall. The three, who usually worked somewhere on 500 Level, had gone directly to No. 6 borehole to work on the hangup that had refused to move for two shifts. They hosed the bottom of the muck column once and had even tried dynamite, but the stubborn hangup refused to move. Hazelhurst, one of the best loader operators Hecla had, had made about six passes with the ST-8, and at 10:00 a.m. had had the loader in the crosscut in front of the shaft with the engine shut down. The wetting was continued and finally the muck began to move, slowly at first, then a little more, until it became obvious that they had succeeded in loosening the entire enormous column of drill cuttings that rose eight hundred feet within the confines of the shaft.

But this time, the flow of drill cuttings did not stop where it always had in the past, at the base of the borehole. Retreating before the awesome mass of moving red muck, Deeder and Udall ran from the crosscut and around the corner of 500 North, toward the dead end of the drift. At the same time, Hazelhurst jumped on the ST-8 and, after several attempts, managed to start the machine, but could not control it as the tremendous flow of muck began pushing the twenty-eight-ton loader out of the crosscut. The ST-8, with Hazelhurst on it, was pushed out into 500 North, where it came to rest against the far rib and the three pipes that carried the compressed air, the water, and sump return. Hazelhurst had absolutely no choice but to abandon the ST-8 or be crushed in the operator's seat. As he jumped from the machine, the onrushing wall of muck fell against the back of his legs, but he was able to continue his flight down the drift. In seconds it was all over. Hazelhurst ran up 500 North, while Deeder and Udall must have surveyed the bleak situation they were suddenly in.

Through circumstance, or the luck of the draw, two of the three men wound up on the dead-end side of an enormous muck run. The third man, probably because he was the operator and the seat of the ST-8 is on the left side, was in a position to run naturally to the south, or to the open end of 500 North. To talk about where the men should have been standing or where the loader should have been parked is so much speculation. The fact of the matter was that no one, but no one, even remotely believed that such an uncontrolled muck run was possible. Hadn't the technique of bottom mucking a shaft been standard for months? Wasn't the practice accepted as standard in mines all over the West? How could something supposedly so normal, so routine, so everyday, go suddenly awry?

Doubtlessly, those are exactly the questions that David Deeder and Terry Udall asked themselves and each other as they assessed their situation in the dead end of 500 North. If there is ever a feeling of total helplessness, that must be it, for there is nothing you can do. Nothing. You wait. In the darkness and the silence and the slowly building heat. You might tap on a pipe, hoping that your friends are a short distance away and preparing to get you out. But mostly you wait. And think. And that must be the worst part of it. All you know is that every effort, every human effort possible, will be made on the other side of that wall of muck to get you out. You know it because, if the tables were reversed, you would be frantically digging in the muck yourself to rescue your friends. Deeder in particular knew that, for at the Sunshine disaster, he personally led a rescue team in an effort to reach ninety-one men trapped underground. Now it was his turn to wait. And to tell Udall that the men were doing everything possible. Maybe in hours, or a shift or two, they would hear the tapping on the pipes and then see the point of a shovel break through at the top of the muck pile that trapped them. Behind that would be the familiar yellow glow of a miner's lamp, and then the sound of human voices. C'mon, Pard, over here. You okay? Good. Let's go, let's get the hell out of here. That was a close one alright. C'mon, we got a jeep waitin', let's get to the surface and get some air. Pard, we lucked out there. To hear words like that would be worth a billion dollars. In the meanwhile, you wait. Conversation? What is there for two men to say? Their whole world has been reduced to the simple passive act of waiting. A few words of encouragement, maybe. And wait. In the darkness, and the silence, and the slowly building heat. Wait. Maybe look at the top of that muckpile where you know the shovel will come through, where you know the miner's lamp will appear and under it will be a filthy, greasy, sweaty face with a grin a mile wide. C'mon Pard, you okay? And wait.

It was just as well that David Deeder and Terry Udall were unable to realize the actual extent of the muck run and just how much rock lay between them and the light, the conversations, and the open drifts that led to the surface and the sun and the blue sky. The cause of this "muck run that couldn't have happened" was a large amount of ground water that had saturated the entire column of rock cuttings that remained in the shaft. Enough ground water had been collecting on top of the cuttings to necessitate the use of a pump, but the amount of water removed was three to five gallons a minute. There was apparently a significant amount of water seeping into the column at lower depths, however, water that was unknown since the bottom of the shaft was dry. The absorption of water within the column of cuttings had created a highly plastic, or fluid, condition. It was not something that had happened instantly, but rather a condition that had occurred over a period of weeks. Just how long it had been waiting to happen is a good question. I recalled the words of the shift boss the previous week: "When she starts comin', there's no tellin' when she'll stop." He had no idea how right he was. Although we didn't know it at the time of the accident, the total muck run that poured out of No. 6 was more than 57,000 cubic feet and weighed about 3,190 tons. In cubic yards, that came to roughly 2,100, an enormous amount considering that it had to be removed eight cubic yards at a time in a run more than half a mile long. The mass of red muck filled the entire crosscut and the main 500 North drift, from the floor to the back, for a distance of nearly 227 feet. Deeder and Udall were trapped behind that wall of muck in a sealed chamber five hundred feet long, and it was calculated that the two men had approximately eighty hours of oxygen in the drift. Although mucking was slow, if the operation could have been sustained at seven feet per hour, it was likely that Deeder and Udall could be reached in about thirty-two hours. It looked like all the rescue would take would be a lot of sweat and effort, if there were no complications. But, as I had mentioned before, if something can possibly go wrong underground, count on it to do so. This time the complications were of the gravest nature.

When Hazelhurst was forced to abandon the loader in the face of a deluge of thousands of tons of muck, the engine had been left running. Turning those diesels off was seldom easy. It was done through a spring loaded fuel cut-off lever, which the operator pulled until the engine quit. Many of those loaders would take five or ten seconds to shut down. The terrifying flow of muck had carried the ST-8 out of the crosscut and against the rib of 500 North. When the loader slammed into the service pipes, a joint ruptured on a six-inch compressed air main. While the engine continued to run, the wave of red muck engulfed the entire loader in seconds and forced itself in both directions down 500 North. The V-10 diesel engine, even though buried, was still able to run, as it was now supplied with air from the ruptured pipe. It was unable to cool itself and apparently overheated until it exploded, igniting the fuel in the tanks. The combination of diesel fuel and compressed air created an effect analogous to that of an oxy-acetylene torch. The resulting fire burned its way upward through the muck, baking it as if it were in a kiln. It melted the cast aluminum engine block and burned through the top of a steel set, indicating temperatures that exceeded 2,000° F.

The odds against this particular set of circumstances occurring were astronomical. To make matters worse, almost directly over the fire, above the steel sets, there was an unusually high back that was packed with timber cribbing. Once that got going, the adjacent timber quickly ignited and the flames were on their way.

The fire burned for two and a half hours, and the smoke and fumes were contained in the mountain of muck. No one was remotely aware of the fire until diesel fumes began emanating from the muck pile. A group of five of us walked on foot to the limit of the muck pile, taking care in the narrow drift to avoid the ST-8s that roared by every few minutes. I remember how totally foreign and out of place that tremendous pile of red muck looked blocking 500 North. Along the rib, the service pipes disappeared into the muck. The air line which Hazelhurst broke in the muck run had been turned off, and the two-inch water line, which was assumed to be intact, was converted to an airline to provide a source of fresh air to the two trapped men. We hammered on the pipes hoping to hear a response, but received no return signal. The huge amount of muck had probably deadened the return signals that Deeder and Udall had probably given. About 12:30 p.m., a hot, acrid gas that made our eyes tear so badly we could hardly see began to seep out of the muck pile. A safety man showed up to check the carbon monoxide content of the air next to the muck pile. He got a reading of 1,500 parts per million, extremely high considering that fifteen times less can kill a man in eight hours. This indicated that a major fire burned somewhere in the muck pile.

The small group of us walked the quarter mile out to the muck bay where the rest of the crews waited. We had no sooner reached the entrance to 500 North when a cloud of smoke and gas, light at first, but rapidly building in density, rolled out of the drift. In only a few seconds, clouds of the thick gray smoke had reduced visibility to six feet. Without any order that I remember being given the miners turned towards the mechanic bays and began walking, simultaneously detaching the new self-rescuers from their belts. The floor of the drift was soon littered with the discarded halves of the outer canisters, and everyone had one of the emergency breathing devices stuck in his mouth. After taking brief refuge in the mechanic bays, which had a bent flow acting to keep the smoke and gas out, we were ordered to evacuate the mine. Visibility was reduced to virtually nothing at the 500 skip station in the waves of billowing smoke. Every man underground had been verbally warned of the fire and all lower levels had been cleared. The vent system was, at that very moment, funneling the smoke and deadly gas to the lower reaches of the declines. When Yates had driven to the South decline heading to pick us up, it was because he needed operators, not because he knew that a fire existed. I wondered what would have happened if we had remained there until the smoke and gas rolled out of the gaping end of the vent bag. Through a lack of communication, the stench warning system was never even activated. When the skip finally arrived at the 500 station, it was possible to see another face only if it were closer than a yard away, and the grimy skin, tearing eyes, and mouths distorted with the mouthpiece

Gearing up for the shift in the mine dry. (ASARCO Incorporated)

of the absurd self-rescuers combined to produce a picture that was simultaneously comical and pitiful.

The self-rescuers had been issued only four days earlier. The timing had been bizarre and had involved some monumental luck. If there had been a lot of joking on August 13 about the clumsy canisters we were given, there was none on August 17, for those much-cursed and verbally abused self-rescuers saved many lives that day. Only through their use did 110 of the 112 men who went underground that day at the start of the shift reach the surface safely. Only two men, Deeder and Udall, remained underground, trapped by 225 feet of muck with a fire in the middle.

I have worked through a lot of shifts when it really felt good to come out of a mine, but nothing approached the sensation of clearing the portal this time and seeing the brightness of the summer sun. As the clean surface air hit their sweaty faces, the crew removed the self-rescuers, bringing with them long strands of thick saliva. For once the hot desert air felt cool and refreshing compared to the scorching air we had been breathing through the self-rescuers. As the crew walked slowly and silently toward the dry, thoughts were about the two men who were still down there. Brass was turned in and counted to confirm that there were really only two men trapped and that everyone else had gotten out. Every man knew that it could just as easily have been he trapped in the end of 500 North. Once again, I conjured up a mental image of the balance that always recurred to me when a man was hurt or killed; the pan on the left with the paycheck and the pan on the right with the slab of rock. In this case the pan on the right was an enormous muck run that couldn't have happened. Miners are the perpetual optimists. They know for certain that these accidents will always strike the other guy, but when it comes close it becomes something you can feel, not merely a remote intangible possibility that one puts out of his mind, but a very distinct reality. And suddenly the fact that you have been trading this numerical probability for a paycheck can make life seem precariously balanced, almost cheap. Life, one would think, should be a commodity to be traded or risked only for highly idealistic causes. In such noble transactions are martyrs made. But there are no martyrs in the mines, no heroes, and on this day there were only two very common men who wanted nothing more than to draw a paycheck, but who instead drew only the short straws.

The crews gathered at the doors of the dry, facing the portals about two hundred feet away, watching the ominous swirls of gray smoke drift upward from the concrete entrance to the underground. What little conversation there was was conducted in sentences even more clipped than usual.

"Now what?"

"Up to them, I guess," gesturing at the mine offices. "They get paid a lot more than we do to figure out something."

"They best figure quick. It's going to get hot down there, awful hot."

"No vent back there either. Gonna get to 120 fast."

"At least that. God knows what that fire's gonna do."

"Jesus Christ."

"Well, what the hell are we gonna do about it? Stand here and look at it?"

"Up to them." Another gesture at the mine offices where the decisions had to be made. "They got their hands full now."

Indeed they did. During the evacuation, representatives of both the State and U.S. Bureau of Mines had been summoned from Phoenix, and both had reached the mine in the early afternoon. A formal closure order was issued, since smoke drifted from both portals and the carbon monoxide levels were lethal. The order read in part:

> *Upon making an inspection of this mine, . . . the underground area is such that a danger exists that could reasonably be expected to cause death or serious physical harm immediately or before the imminence of such danger can be eliminated. That condition being a mine fire with high CO readings. You are hereby ordered to cause all persons except those necessary to perform the work in order to comply with this Order to be withdrawn from and to be debarred from entering the Lakeshore Mine.*

Considering the conditions, the order was academic.

At 3:45 p.m., a second rescue effort began when a small team wearing self-contained oxygen breathing apparatus (OBA) went underground. Utilizing the bulky oxygen tanks and full face masks, the men reached the fire area and turned on the air vent fan to force air into the forty-eight-inch vent bag leading into 500 North, hoping this would clear the drift of the dense smoke. The six-inch compressed air main was left off in the belief it was providing oxygen to the fire. All these actions were speculative only, since the exact nature of the fire was not known at the time, nor was the true condition of the pipes. It was, in essence, a period of blind experimentation.

There were few courses of action to choose from at this point. Basically, they came down to attempting to supply the trapped men with air, or attempting to put out the fire. To do one would in all probability be detrimental to the other. The decision had to be based on very sketchy and incomplete facts and upon educated guesses and suppositions; whoever made the decisions would also have to live with them when it was all over. It was one thing to stand in the dry looking at the smoke drifting lazily out of the portals and venture all sorts of suggestions and courses of action to take. But every miner also knew that it was a different story to make and be responsible for the decisions when they were finally made.

The hope was that, without the air line, the fire would burn itself out. At 6:00 p.m. another team donned the OBA and went underground to report the smoke was diminishing. Succeeding crews advanced the vent bag forward to establish a fresh air base near the accident area. Throughout the evening, the levels of carbon monoxide in the exhaust air at the portals slowly diminished, and it appeared the fire was dying out. Thoughts turned from the fire to the concerted mucking effort which would be the next step. Optimism was at its highest in the evening, and it looked like Deeder and Udall might be out in about forty hours if everything went well. But the optimism didn't last long, for it was soon proved to be wishful thinking. About midnight, some fourteen hours after the accident, the carbon monoxide levels outside the portals suddenly soared, this time to their highest levels yet, a certain indication that the fire was nowhere near out as hoped, and might even be worsening. In the early morning hours, the newly rigged air vent bag was turned off, as it was assumed to be the source of oxygen for the continuing fire. It had now become obvious that it was much more than a simple matter of mucking to free the trapped men.

On Saturday morning, in the pre-dawn darkness, sixteen miners trained in mine rescue techniques arrived from Magma Copper Company's San Manuel Division to bolster the force of forty-one Hecla men who had received similar training. It was decided that, in order to reach the accident area to extinguish the fire, a series of portable bulkheads spaced a few hundred feet apart would have to be erected to advance the fresh air base and thus, in a pattern not unlike a military operation, reclaim step by step the 1,300 feet of smoke-filled drift leading to the muck pile. This was hard, heavy work, and it had to be performed while wearing the clumsy OBA in temperatures ranging from 118 to 135° F. where the humidity was one hundred percent and the carbon monoxide level was three thousand parts per million. Later in the morning of that second day, three men arrived from the National Mine Service with additional OBA equipment, while two men from Centennial Development Co. and three more from Cementation Company of America, both Hecla contractors, also joined the rescue crews. As more trained men became available, eleven mine rescue teams were organized and scheduled in round-the-clock operations to perform the slow, laborious job of advancing the bulkheads into 500 North.

Also on Saturday morning, while the rescue crews crept slowly forward, and while David Deeder and Terry Udall had completed their first twenty-four hours of entrapment, the *Phoenix Republic* hit the street with the first printed news of the underground accident. The headlines of every paper in the country, in August of 1973, dealt with Nixon's fight to retain possession of his famed tapes. That was the headline again on August 18, but midway down the first page appeared:

TWO HECLA MINERS TRAPPED AT 1,200 FEET

The article which followed was a short account in the most general of terms. It stated little more than that two men were trapped by a freak accident in an underground mine. This was the start of the steady press coverage of the unfolding drama of the rescue efforts at Hecla. Tragedies such as underground mine entrapment have always made good copy because of a basic, morbid fascination on the part of the general public. Death, whether by car accident or by something a bit more exotic and irregular, is exactly the same. Death. But it is the means that make the difference. The story of a death by car accident or natural causes is so readily comprehensible as to be mundane. But to the average person on the street, even the simplest workings of an underground mine are shrouded in mystery, all a part of a foreign world one reads about on infrequent occasion, a dark, eerie nonreality that exists in inaccessible places deep in the earth. Entrapment under those alien, hostile, and little-understood conditions then becomes an exaggerated, symbolic drama of human life against awesome, dark forces that are the embodiment of evil. It is the ultimate stage play for which the public is the audience. Any two men critically injured in an industrial accident on the surface and barely staving off death in a hospital bed would warrant a second- or third-page blip in any newspaper, and that would be it. Yet two men trapped in a mine 1,200 feet beneath the surface and staving off death in a different manner will get the full press treatment. The newspapers and other media realize fully that there is a repulsive mystique surrounding underground mining that provides a solid foundation for a natural drama.

Meanwhile, the underground rescue crews made slow but steady progress in advancing the bulkheads toward the muck pile. The assumption had apparently been made that once the men did reach the fire, it would be easily extinguished. Hence, no other attempt to reach the trapped miners had been initiated.

On Sunday morning the headline of the front page article read:
TWO TRAPPED MINERS BELIEVED STILL ALIVE

Information about the rescue attempt was rigidly controlled by the company and the resulting articles built a facade of optimism. This article cited the natural air supply that was contained in the sealed drift five hundred feet long where the men were trapped. Arizona State Mine Inspector Verne McCutchan carefully explained that the miners "were trapped, not buried," and that the lack of response to the pipe signals was because "the muck pile is long and could deaden the sounds of the tapping." The remainder of the article went on to explain that the mine had struck "hard luck" and, after experiencing repeated delays, was two years behind schedule.

In the light of the local and national media coverage, images were also at stake. This was true not only for the Hecla Mining Company, but also for the state mine inspector, which, incidentally, is an elective office in Arizona. From his standpoint, it would be awkward, now that an accident had already occurred, to criticize a mining company for having poor safety standards. This official was himself responsible for enforcing the safety regulations. Accordingly, the wording of any and all press releases was given the utmost attention, and the assistant to the president of the Hecla Mining Company himself had arrived at Lakeshore to supervise the releases. Certainly, if a totally confident posture could be maintained during the rescue effort, and if the trapped men were brought out alive, it would look very good for everyone concerned. There are secondary things of some importance to consider aside from just getting two men out of the hole alive.

On Monday, August 20, the front-page article of the *Phoenix Republic* announced:
EFFORT TO RESCUE TWO TRAPPED MINERS
MOVING TOWARDS FINAL STAGE

The term "final stage" was absurdly premature. The article was concerned mostly with the alternating streams of air and water that were being sent through the two-inch pipe into the muck pile, and, it was hoped, into the sealed dead-end drift. The article was continued on page four, and beneath it was a related article headlined:
STATE MINE INSPECTOR CALLS SENATOR A POLECAT

A lot of miners were disgusted by this intrusion of cheap politics into the rescue effort, but politicians, recognizing a good sounding board when they see one, rarely pass up the opportunity. Arizona State Senator John Scott Ulm, seeing an outstanding chance to get his name in print and express his overwhelming concern for the welfare of poor miners everywhere, called State Mine Inspector Verne McCutchan "an apologist for the mines who had not carried out safety inspections." The fact of the matter was that the muck run and fire had nothing to do with mine inspections and safety standards and was, indeed, a freak accident. The general public certainly wouldn't know this and I doubt whether Ulm did either, but it was an effective smear on McCutchan and made Ulm look as though he were really concerned.

During the rescue operations only one crew at a time could work in the narrow 500 North drift, meaning the rest of the men rested on the surface. There was a lot of card playing going on to pass the time between shifts, and McCutchan was known to have played his share. A man can only check on so much progress, set up so many schedules, and write so many reports, which McCutchan apparently did, and the card playing was only an effective and necessary means to pass the long hours. Standing at the portal with clasped hands muttering prayers in a long vigil wouldn't have accomplished a damned thing. Ulm, however, seized this opportunity to smear McCutchan and to imply that he was somehow to blame for the accident. It was unfortunate that McCutchan could not have ignored Ulm's remarks as though they were unworthy of reply. Instead, he chose to fight fire with fire, and launched his own smear of Ulm. He called Ulm a "disgrace to the Senate and the news media," and brought up a bad-check charge in another State that had caused Ulm to seek a pardon from the governor of Arizona before he could legally run for public office. So while the public was reading about the childish exchanges of the two bureaucrats shouting slurs at each other, the rescue crews were trying their best, and David Deeder and Terry Udall, if they were still alive, were completing their third day of entrapment underground.

The rescue teams succeeded in reaching the fire area at 10:00 a.m. on Monday. Glowing embers were seen in the timbers, but, since this was now an oxygen-deficient environment, no smoke was observed. The temperature was 133° and the carbon monoxide level still hovered around eight hundred parts per million. After completion of the final bulkhead, the crews used fire hoses with fog nozzles to wet down the fire area until a foam machine was set up to pump a fire-retardant foam. The foam machine was used all night, but did not succeed in extinguishing the flames. Finally, when it became apparent that the fire would not extinguish easily, attention was turned at last to the possibility of drilling a borehole from the surface into the entrapment area. It was noon Monday before the decision was made to locate drill rigs in the area capable of sinking a borehole the required 1,200 feet. The drilling of a 6½" borehole was started from the surface at 9:00 p.m. on August 20, 3½ days after the accident. The purpose of a successful borehole, of course, would be to lower a communications

A mine rescue team wearing self-contained oxygen breathing apparatus (OBA). Crews wore identical equipment during rescue operations at Hecla's Lakeshore Project. (ASARCO Incorporated)

probe and to supply the men with food and water. One would think that such a borehole would have been started within hours after it was realized that a fire would complicate the rescue. Why was the drilling not started, or even ordered, until three full days after the accident? Good question. Apparently because the men who made the decisions assumed from the start that the fire could be extinguished quickly and the mucking operation could proceed. The theoretical eighty hours of air that the two trapped men could rely on had expired about the time the drilling operation was begun.

Tuesday's front page article headline read:

SHAFTS DRILLED TO AID MINERS

This article recounted the commencement of the borehole work the previous evening and pointed out that the rescue crews, using the OBA with its two-hour supply of oxygen could work only thirty minutes at the accident site. "The men," mine officials said, "have an excellent chance to be alive because it is believed they escaped the slide and are being supplied with water and air through a two-inch pipe."

At this point, many of the rescue team members themselves did not think that Deeder and Udall were still alive, and they stated that belief when questioned specifically about it by mine officials at the ends of their shifts. Furthermore, Ed Hazelhurst, who was being subjected to considerable pressure by management for his part in the accident, and who was driving himself as hard as he could in the rescue work, noticed that a valve on the two-inch line—the same pipe that was supposedly supplying the trapped men with air and water—had been *shut* all along. At the end of his shift, he reported this to management and was ordered to keep this information quiet for fear of demoralizing the other men. The members of the rescue teams themselves, who could have given honest, firsthand accounts of what the conditions were really like underground, were barred from speaking with the press.

On Tuesday, the crews advanced past the last bulkhead directly into the fire area. They found that the foam that had been pumped into the fire had advanced only a mere forty feet, and that the remaining 210 feet had been unaffected. Virtually all the timber had been completely burned and all that remained were the steel sets, some of which had been severely damaged by the intense heat. At this time, the temperature in the drift was 143° Fire hoses were again brought in and used to wet the entire area down for twenty hours. It was during this time that the fire was finally extinguished. If the rescue teams had received any one break during all this time, it was that the intense heat had somehow failed to create any new rockfalls.

The press optimism continued on Wednesday with still more front-page coverage:

FLAMES DOUSED AND HOPES RISE FOR TWO TRAPPED MINERS

This article made heavy mention of the report that no new rockfalls had been encountered. Hecla officials must have thanked heaven for some honest good news to give the papers. At the time the public was reading about all the "rising hopes," Deeder and Udall had been trapped for five full days under conditions which could only be imagined. There is a big difference between honest hope and self-delusion, and, no matter what the official, carefully worded releases said, every man there knew it was 143° on one side of that muck pile. Only a damned fool could expect it to be significantly cooler on the other side. As a final bit of irony, the continuation of the Wednesday article on an inside page was directly adjacent to a picture filler of a girl in a bikini splashing around in the California surf. It was titled "Beating the Heat."

On Thursday, when the entrapment had stretched to six days, the news accounts in the *Phoenix Republic* had been relegated to the first page of the second section:

RESCUE OF TRAPPED MINERS BELIEVED POSSIBLE TODAY

Although the newsworthiness of the rescue effort may have diminished somewhat in the eyes of the newspaper editors, the official optimism did not diminish one iota. The article said that all-night fogging had reduced the drift temperature from nearly 150° to 110°, another "encouraging" factor. Such undying optimism has a dual purpose. First is the public image of the company itself, the public's association with the good company name, with the driving effort that never stops, with the official belief that, no matter what the odds, they're going to get their men back, and with the human concern that is the overriding factor in all decisions. Second is the fact that the closure order would be in effect for the entire mine until the rescue efforts had been completed, and any work done must, by law, be related to the rescue. In other words, the mine development had come to a halt. Another delay. And in mine development, time is money. 500 North was going to have to be cleared sooner or later, and it may as well be sooner with the round-the-clock work of the rescue crews. In the light of the official optimism, the crews would have little choice but to continue their exhausting and relentless effort, even though few, if any, really believed the men were alive after six days of entrapment.

As the public read that the rescue was believed possible that day, the surface drill rig, after sixty hours of work involving many delays, had penetrated only 648 feet, a little more than half way to the dead-end drift.

On Thursday afternoon, the mucking had cleared the drift enough to reach the burned ST-8 that was buried opposite the cross-cut. In the process, another ST-8 used by the rescue crews caught fire because of an electrical short circuit. But the opreator, as he abandoned the machine, was able to activate the remotely controlled fire system, which extinguished the flames. After the buried loader was uncovered, there was not enough room to move a Second ST-8 past it in the narrow drift. Rock bolts were used to support the back, and two steel posts were removed on the opposite side of the drift to allow room for a much smaller loader, an ST-2B, to squeeze past and continue mucking.

Seven full days after the accident, on Friday, the *Phoenix Republic* brought the rescue story back to the front page:

MINER RESCUE TEAM DIGS BY HAND

That apparently referred to the efforts to bypass the burned-out ST-8 that partially blocked the drift. The assistant to the president of Hecla threw in the comment that "the men were working with renewed encouragement," but he no longer offered any estimate, as he had done previously, as to when the breakthrough would happen. Also revealed, doubtlessly through reporters' questions, was the fact that the Hecla Mining Company owned a share of the Sunshine mine in Kellogg, Idaho, where, the article continued for the sake of its non-mining-oriented readers, ninety-one men had died in a fire a year and four months earlier. The reporters, knowing a good angle when they saw one, dug further and elaborated on another Sunshine connection. Deeder had led a Sunshine rescue crew, and even Terry Udall had done some mining in Kellogg. The article closed with the U.S. Bureau of Mines' stated findings on the Sunshine disaster, which were released only six months before. The report blamed the company in large part for the disaster, which the findings said had been caused by violation of certain safety standards.

The mucking with the ST-2B continued into Saturday, and it was becoming obvious that the breakthrough would be soon. The front-page article headline on Saturday, August 25, was:

BREAKTHROUGH REPORTED NEAR AT COPPER MINE

For the first time, the headline was a simple objective statement devoid of all the previous optimistic trappings. A breakthrough was near, and that was that.

While the ST-2B continued to eat into the muck pile, the borehole overhead had penetrated 1,244 feet by 7:00 a.m. and had broken through the back of the dead-end chamber. A communications probe was lowered but hung on an obstacle at 1,140 feet. Once again the drill stem was inserted to clear the hole and touched the floor of 500

North. After the drill pipe was reported clear, the probe was lowered again and came to rest atop a piece of cribbing in the sealed drift. At 4:40 p.m., a communications technician on the surface begain listening for any sounds which might indicate life. He heard none.

At 5:00 p.m., eight days and seven hours after the muck slide had trapped David Deeder and Terry Udall, the breakthrough was achieved. On the surface, Hecla officials had allowed TV newsmen to set up lights near the portal to film the moment when the trapped men were brought out. An hour later a recovery crew entered the sealed drift and found the bodies of Deeder and Udall. The temperature was 120° and the carbon monoxide concentration was one hundred parts per million, still a lethal level. On the surface, the cameramen and newsmen were ordered back to a parking lot while the bodies were brought out. The rescue crews were released to go home after eight days of monumental effort, the end result of which was, unfortunately, recovery of two bodies and clearing of the 500 North drift. When it was determined that there were no harmful levels of carbon monoxide remaining in the mine, the closure order was lifted.

On Sunday, the *Phoenix Republic* concluded its front page coverage of the eight-day rescue attempt:

TWO MINERS ENTOMBED FOR WEEK FOUND DEAD

It was a statement of utter finality. It showed the hollowness of the previous week's headlines and the futility of the whole thing. Quiet prevailed at Lakeshore at last, and the crews were not required to return to a regular work schedule until Tuesday. All that was left now was some mopping up on the part of the press, some paperwork, and the inevitable verbal exchanges between involved persons.

The initial autopsy conducted by the Pinal County Medical Examiner indicated that the two miners died "as a result of the combined effects of dehydration, heat exhaustion, and carbon monoxide intoxication." Death had occurred, according to the report, five to eight days prior to the examination.

During the next week, various articles appeared in the *Phoenix Republic* mentioning items that were not available to the press earlier. One carried the story of Hazelhurst's report that the two-inch pipe, upon which rested the hopes of air and water reaching the trapped miners, had been shut off for the first three or four days, and that Hecla officials had ordered this information kept confidential. McCutchan countered this with an argument that Hazelhurst "must have been confused" and that he was "certain" the trapped men were receiving air and water. The mine inspector went a little further to say adamantly that the men would have survived if they had remained near the pipe at the far end of the sealed drift.

Another article appeared stating that Hecla had not announced that dynamite had been used to blast the muck loose on the morning of the accident. In the earlier official news releases, the company had said that the muck run started inexplicably after the operator had made one pass with the loader. Hecla had refused to allow reporters to interview Hazelhurst, but the truth of the matter was that the three miners went to considerable lengths, including six passes with the loader, extensive hosing, and the use of dynamite. Inspector McCutchan, when asked about the use of the dynamite, was later quoted as saying, "I was aware of it, but it didn't seem important enough to announce. You know, what seems important to the press may not necessarily seem important to miners who do that sort of stuff everyday." Actually, it didn't matter whether dynamite was used or not, or how many passes were made with the loader; both were standard underground procedures and in no way violated any safety regulations. Those three men were sent into No. 6 with specific instructions to get that muck moving, which, ironically, is just what they did.

The muck slide was, in all honesty, a freak accident and didn't require a lot of public relations special handling. In fact, an honest approach would have contributed a touch of dignity to the deaths of David Deeder and Terry Udall. In a way it was a shame that the press was unable to interview the members of the rescue crews. Doubtlessly they would have been able to print some accurate and striking accounts of what it was really like in the smoke and heat of 500 North, instead of the bland, optimistic generalities they did report. Since it is a rare reporter who has a realistic idea of what normal underground conditions are really like, much less a fire and cave-in situation, the company was in a commanding position when it came to press questions. And when it was all said and done, the public, even with all the attendant publicity surrounding the tragedy, still had no real idea what the world of the underground miner was all about.

And for a final exercise in futility, consider this paragraph from the U.S. Bureau of Mines Health and Safety Report. The Fatal Report of Entrapment of Men and Mine Fire, Lakeshore Mine, Hecla Mining Company, Casa Grande, Pinal County, Arizona, August 17, 1973:

> . . . *Based upon concentrations of carbon monoxide which were known to exist on the south side of the muck pile during the firefighting operations and upon the fact later investigation revealed the 6-inch air line and the 2-inch converted water line were ruptured at the loader, thus providing an open channel for carbon monoxide to enter the otherwise sealed area, it is probable that the trapped men died of carbon monoxide poisoning within a few hours of their entrapment* . . .

The last week in August was still hotter than hell, with temperatures regularly around 105° and 110°. It was good weather for Mexico and the Gulf of California, where the deep blue shimmering water and cold beer could make all of what had happened over the past week and a half seem as though it never had happened at all. But it did, and David Deeder and Terry Udall would not be making the next shift. I remembered Udall's comment after getting back from a Mexican weekend that it always seemed he started more shifts than he quit. In the end, that proved very true. He had started one more than he had quit.

The crews were noticeably subdued as they drifted back together again after a few days off. Some of the overhead baskets in the dry that had been full before the accident were now empty, mute testimony to the tramping that had begun. At one time or another every man there probably entertained the idea that just maybe it was time to move on, to look again for that non-existent mine where such accidents didn't happen and where conditions in general would be a lot better than they were in this rotten hole. Not much was said, but the nods of greeting and recognition all had that same peculiar "well, here we go again" look.

New self-rescuers, bright and shiny, were issued to the entire shift to replace the ones that had so suddenly and unexpectedly been expended. There were no jokes this time as the miners strapped the canisters on their belts and headed out for the familiar walk to the portals to wait for the skip. The declines, as well as the rest of the mine, were thoroughly permeated by a heavy scorched stench that served as a reminder of the fire and the long fight to put it out. Still littering the drift near the entrance to 500 North were the self-rescuer canisters that the crews had discarded as the clouds of smoke and gas broke through the muck pile and rolled down the drifts.

There was no official push to get started where we had left off, and it is unlikely that a request for fire and smoke at quittin' time would have done any good. There was a lot of cleanup and repair work to be done, and the decline headings had filled with nearly one hundred feet of water during the long period when the pumps had gone unattended. There seemed to be an unspoken agreement that work would proceed slowly, and that none of the deadline push that had been a daily dilemma before the accident would be necessary. Lunches lingered well past the allotted half hour everywhere.

"Well, quarter to. Think we better start movin'?"

"Another ten minutes, Pard. They ain't gonna say nothin'."

The small groups of five or six miners would remain sprawled on the timber and muck piles. Most of the joking and horseplay were absent.

In the darkness, beneath the omnipresent cap-lamps, are human beings. (ASARCO Incorporated)

"When do you think that goddamn stink will quit?"

"That'll be here awhile, it's all over the mine. S'matter, you don't like it?"

"Makes me think of the fire."

"Yeah. Been back in 500 North yet?"

"Once. Bad back there. Gonna take a long time to clean that mess up. When are they gonna haul out what's left of that loader?"

"Don't know. Don't much care, either."

The staccato conversations would lapse back into silence, and only the throb of the pumps would echo along the drifts. The shifts were easy now, just a lot of cleaning up and killing time, and each man would be content to lie back and be alone with his own thoughts, most of which dwelled on better and more worthwhile things that would start only when the skip reached the surface. Occasionally, the question which was often thought but rarely asked would interrupt the silence.

"How long do you really figure those guys lived?"

No answer. Only the throb of the pumps.

To say 500 North was a mess was an understatement. The drift was filled with blackened timber and ash, twisted steel that the intense heat had melted, pieces of ripped vent bag, puddles of water in the rutted drift floor, huge quantities of the red drill cuttings packed into the ribs, and worst of all, that acrid stench. Along with two other miners, I was sent back into 500 North to begin replacing the steel sets and the timber that had burned away in the fire. The ST-8 which Ed Hazelhurst had abandoned two weeks ago in the face of the deluge of muck still rested against the rib.

"Hey, did you look at that engine?"

"Something, ain't it? Took a lot of heat."

"That's it, ain't it?"

"All that's left."

The loader was still caked with baked red muck, and all that remained of the massive ten-cylinder aluminum engine block was a piece of blackened slag about the size of a basketball. Even though the heat was tremendous, the engulfing wet muck had insulated most of the loader, and fire damage was centered on the engine compartment. Portions of the six-foot-high tires had been burned away to reveal the steel cords. Much of the metal paneling on the loader was crushed inward, confirmation of the thousands of tons of muck that had poured out of No. 6. The crosscut itself was still completely packed with the red muck.

"Okay, you guys know what you're supposed to get done, right?" Luke Byrd asked.

"Yeah, we'll get 'em up."

"Now be careful and remember," he added, gesturing with his lamp beam through the darkness. "Nobody goes up there. Nobody."

We all turned to look at the small doghole at the top of the remaining muck pile, our lamp beams dancing along the slope of red cuttings only to get lost in the pitch blackness beyond where Deeder and Udall had put in their last, long shift.

"And another thing," he said as he climbed in the jeep, "if any of you hear anything comin' out of there," motioning to the left, towards the mass of red muck that still encased the bottom of No. 6, "or even think you hear anything . . . well, get the hell out of here quick."

"Just don't get in our way with that jeep."

Slowly, over the next several days, the essential ground support was replaced and the mangled, burned-out loader was hauled up to the surface where the surface crews could only stand and stare at it and wonder what the hell had happened down there. Another closure order had been issued for 500 North in the area of No. 6. It cited the possibility of another muck slide, the only exception being for the men who had removed the loader. When that job was completed, 500 North was roped off and became simply a dark place where no one went.

The last event dealing with the fatal muck slide was a county inquest held by the Pinal County justice of the peace. Prior to the inquest, the justice of the peace was quoted as saying, "I don't expect anything criminal to come out of this. Mining is a hazardous occupation and it appears that no one is to blame for this. We'll just go through a list of witnesses and they'll testify as to what they know." That comment basically set the tone for the inquest, which was a rote exercise to fulfill the paperwork requirements necessitated by law. The inquest convened in Casa Grande in early September when it was still one hundred degrees.

Just what it accomplished was uncertain. Hazelhurst and members of the rescue teams were questioned by county officials who didn't have the faintest idea what underground mining was all about. They tripped and stumbled through the use of mine nomenclature, unable to differentiate between such basic terms as "drift" and "shaft," often using them interchangeably. The questions asked, through no real fault of the county officials, exhibited such an utter ignorance of mine procedure and environment that the whole thing became almost a farce. But it did serve to satisfy the requirements of law, and Pinal County could rest in peace with the paperwork completed.

By the end of September, most of the summer heat was past, and the mine had gotten back to normal. The scheduled development of the mine was now even further behind as a result of the time lost through the accident. The relaxed attitude that had lasted several weeks was gone, and a bigger push to get things done faster than ever before was felt from the highest offices down to the lowest shifter. The decline crews rarely left the headings now, and all effort was turned to driving the remaining distance to the terminal point at sea level. As the surface temperatures dropped in the fall, the underground environment improved greatly, even though there had been no improvements to the air vent system. The air quality, of course, remained as rotten as ever, but a significant reduction in the underground temperatures made the shifts much more bearable. Sometimes, on graveyard, the storm of air that would scream down the two vent shafts onto the 500 Level could actually be considered cold, particularly to the operators who would sit for hours on their ST-8s.

To replace the burned-out loader, a brand new unit arrived, was checked out, and was sent into the declines. After one week underground, it was hard to distinguish the new machine from the rest of the ST-8s. The bright new yellow paint was now covered with an inch of grease, muck, and cement, the bucket was scraped and packed with muck, the smooth metal panels were already dented, and a scattering of sunflower seeds covered the instruments.

The declines ran into a little more bad rock and still more time was lost when the intersection with the lowest crosscut had to be poured with cement to keep the ribs from collapsing. An operation like this can turn into a real circus, but this went rather smoothly and safely, considering what was involved. Steel plate was welded onto the posts to form the ribs. The welding was done by a crew running around with high-amperage equipment in knee-deep water. ST-8s would pick up a bucket load of cement from a service shaft on 500 when the operators told the anonymous voice on the phone that the loader was positioned. In a few minutes, a frightful roar above in the shaft told him that the load of cement was on the way, falling 1,200 feet from the surface. Twelve tons of that slop would slam into the bucket, shaking the whole loader and splashing the operator and everything else in the vicinity. There was one loader that saw little but steady cement hauling duty, and with the inch of cement that encased it, one would think the machine was made of cement. The operators would then take their load of cement down the declines in low gear. Even with the steady braking effect of low gear, it took a lot of brakes to stop it where you wanted it in the heading. Then, with an accomplished manipulation of the bucket, the operator would neatly deposit the twelve tons of cement into an air "pot," a steel container which would pneumatically pump the cement into the forms. If anything went wrong, or if an operator exercised a bit of bad judgement, the floor of the drift would be a foot deep in cement and the operator would be the subject of obscene verbal abuse from the miners who were standing in it.

If there was one job worse than hauling cement down the declines, it was working with the shotcrete crew. Shotcrete is a cement-like material that is sprayed pneumatically on rock surfaces to provide a form of ground support. Hecla had a liking for shotcrete, and often tended to overdo a good thing. In places the shotcrete would be so thick on the ribs and back that it would crack and present more of a hazard to miners than the rock itself. Straight work on the shotcrete crew was tantamount to penal duty. The confined areas where the work was being performed usually had poor ventilation, and respirators were an absolute necessity, especially for the poor hand who had the misfortune to be strapped to the nozzle which sprayed the caustic glop with one hundred pounds of air pressure. The shotcrete mist itself could cut visibility to five or ten feet, and, if the crew were working from a bucket, the big diesel engine of the loader would contribute its share of fumes into the air everyone breathed. The shotcrete was a severe skin irritant, and some of the poor guys on that crew looked like lepers.

As the days of fall turned into December, the idea of quitting Hecla got stronger and stronger, and soon became my preoccupying thought.

"Pard," I said to my shifter one day, "think I'm going to be tramping this hole next week."

He stared at me with a look of disbelief. "You serious?"

"Damn straight I'm serious."

"Well, what's the problem, Pard? Anything wrong?"

"No problem," I said. "You know how you can feel like moving sometimes? Well, I feel like moving."

"Pard, I think you're makin' a mistake. I mean this is going to be a good place to work once it gets straightened out. You already been here about, oh, six months, I guess, and you know what that means. Seniority, Pard. Yessir. They'll only be hirin' more hands and you got a chance to be on the bottom floor. And you know what that means."

"What's it mean?"

"Well," as though it should be obvious, "Things could get pretty good for you."

"Things have already gotten good for me. I came here to make some money and I made it. Things ain't about to get any better. You have my notice, right?"

For the next week I received some unusually fine treatment. A lot of good "sitting" jobs where you take a loader somewhere and sit on it. If this keeps up, I thought, I just might hang around a little longer. Discussions during the last couple of lunches dealt almost exclusively with tramping. It is an easy idea to instill in the mind of a miner.

"Hear you're thinkin' of leavin' us, Pard."

"Yeah, pretty soon."

"Where you goin'?"

"No place special, just going. Little traveling, maybe."

"Minin'?"

"Nope. Had enough of that for a while. Might go back down the Caribbean and do a little more salvage work," I told them. I had often told the crew stories about the shallow-water scuba work I had done, and I never failed to be amused by their reactions.

"Divin'?" Wide eyes. "You want more of that? Goddamn, you got more balls than me, Pard. This whole crew couldn't drag me into that goddamn water down there. That crap's just too dangerous."

Unanimous nods of agreement. Everyone of these guys honestly believed that scuba diving presented far more dangers than underground mining.

"Trouble with you guys is that you watch too much of that bull on TV," I said.

"Pard, you've got a screw loose. You're just askin' for trouble. All them goddamn things with teeth swimmin' around down there. Here," gesturing at the floor of the drift, "A man knows what he's got. Yeah, you can get hurt, but you know what to watch out for and how to handle it. Ain't nothin' goin' to come runnin' out of the rib and eat you, for crissake. Not like that divin' of yours."

"I'll tell you guys one more time," I argued. "Your trouble is that you've been down here so long that you've just lost track of how bad these holes are. You think this is a normal job now, that the whole world is like this."

That would always bring a few moments of thought and silence that would be followed by the inevitable comment, spoken with deep sincerity.

"Goddamn, I wish I could get out of this sonuvabitch."

"All you got to do is walk up and tell 'em to stick it in their asses."

"It ain't quite that easy when you got an old lady and a bunch of kids."

"No, I guess not."

"I'll tell you, Pard, when you get married, make damn sure you ain't workin' in a goddamn mine. Easy to come down here, but it ain't so easy gettin' out. If I was single, I'd walk out of this sonuva-bitch right now and tell 'em all to go screw themselves." A pause and a sarcastic laugh. "But that ain't gonna happen 'cause I'm stuck down here and I know it. I can make money alright, but I can't get out."

There was a lot of truth to that. Mining can be a one-way street with precious few turnoffs. It prepares you for one thing, and that is more mining. So many miners adopt a standard of living that eats the entire paycheck, and to get out, to work on the surface, is going to mean a cut in pay and, accordingly, a drop in the standard of living. One can easily get accustomed to a good standard of living.

"When you trampin' out?"

"Whenever the spirit moves me. Maybe when something rubs me wrong."

"If it was me, I'd be brassin' out now."

I smiled, enjoying the pleasure and satisfaction that comes with tramping, knowing you don't have to run to another hole tomorrow. I didn't have to wait long to be rubbed wrong, for the second half of the shift got pretty rotten, and I made up my mind right then and there that I had had it. The idea of getting hurt only a few days before tramping was positively unpleasant. I thought of Udall and Deeder, who never finished their last shift. Well, this was my last and I had every intention of turning in my own brass.

For the first time at Hecla, I was not obsessed with rushing out of the hole as soon as possible at the end of the shift. The skip ride up the North decline could have taken an hour for all I cared. The rushing air felt good, and I was even pleasantly aware of that characteristic mine smell, although it was still tainted with the acrid stench that was a lingering reminder of the muck run and fire five months before.

When I brassed out I told them they could keep it. I took a long, slow shower, cleaned out my basket in the dry, slung my gear over my shoulder, and gave them an address to which they could send my final check. I was nearly an hour behind the rest of the shift, half of whom were already sitting in the Wonder Bar. On the way back along that lonely desert road that crossed the reservation to Casa Grande, I passed a pickup truck stopped by the side of the road. A man and three boys, all Papagos, had filled the bed of the truck with as many Coors cans as it could hold. I doubt whether it had taken them long at all, since the summer had enriched the aluminum deposit sub-stantially. It amused me to think that this Papago was smart enough to let the Anglos kill themselves in an underground copper mine all summer, then use the magnificent days of the southern Arizona winter to run an open-pit aluminum operation.

I pushed open the door of the Wonder Bar and stood there a moment, allowing my eyes and ears to accustom themselves to the gloom and the jukebox. Jeannie Pruitt had finally worn out the satin sheets.

"Hey, Pard," a voice called, "miners only in here."

"I got news for you guys," I said. "I was never a miner in the first place."

"Hell, Pard," came the answer, "wasn't a man on the shift didn't know that all along, but bein' nice guys we let you in anyhow."

CHAPTER V

HIGH PLAINS URANIUM

Long ago and far away, back in the 1960s, I received a grand tour of the Far East, courtesy of the U.S. Government. Today, my memories of that world have somehow evolved into a pleasant composite of the good times, the friends, burning up leave time in Japan, the adventure of hitting the off-limits areas of Seoul, and the general eat-drink-and-be-merry attitude. After all, the Government was paying for most everything. It is only through conscious effort that I am able to recall the dumb, pompous lieutenants, the asinine sergeants, the long nights on the DMZ OPs, the wretched winter field problems, and the alerts that were always scheduled for 4:00 a.m. I had learned that in the military, as in underground mining, the human propensity for selective recollection is enhanced by the passing of the years. In the long run, the whole thing didn't seem bad at all. The eight months I had spent at Hecla had become only a far-off memory of Mexican beer and clear water, paychecks, easy weekend shifts which made the paychecks even better, and a certain secretary who helped me spend them. It was only when I delved into the blacker recesses of my mind, the recesses that I really didn't want to open, that I remembered the heat, the miserable air, the friends that were hauled out dead, and that rock with my name on it.

Contenting myself with the more pleasant memories of hard-rock mining, I spent a year or more roaming around the Caribbean, every day forgetting a little more of just what it took to earn one of those mine paychecks. As I knew it would, the time came when there had to be some checks coming in from somewhere, and pretty quick. What I needed was five thousand dollars in hand, which would be enough to pay another installment on my advance retirement. There were two ways to go about it. First, of course, I could take a nice conventional job and earn the money slowly, possibly saving the five thousand dollars after a year of steady work and austerity, but knowing I would be around to enjoy it when I quit. Secondly, I could go back to the mines, make it quick, and keep my fingers crossed. It had been nearly two years since I had strapped on a lamp, and my thoughts of what underground mining was had been tempered appreciably, although on that skip ride out of Hecla on my last shift, I had told myself, "That's it, Pard, no more." And I had meant it.

Again, it was the classified section of the *Denver Post* in which I sought the source of my five thousand dollars. And there, in the usual big block ad, it stared me in the eye again. "Experienced Underground Miners Wanted Immediately—Near Douglas, Wyoming—Exxon Company, U.S.A." I have to say one thing for mining companies looking for hands—they don't spare the newsprint, no three-liners for them. There was a lengthy list of benefits that would go to the lucky miners who qualified for this one, but I had to chuckle at the "near Douglas." I remembered the "near Casa Grande" only too well. If I could have clearly recalled just how miserable Hecla had been at its worst, I wouldn't have given the ad a second thought. But I couldn't, and there it was all over again. Come one, come all, and seek your fortune down in a nice, cozy mine near Douglas, Wyoming.

Hiring on at Exxon was a more elaborate procedure than that followed by most mines. After filling out the application I had received and returning it, I waited patiently until another letter informed me my presence was requested for an interview at the personnel office at the mine site. I left Leadville, which had become something of a summer rest stop over the years, and headed up I-25 into Wyoming. One hundred and fifty miles across the state line I reached the town of Douglas, an oasis on the vast, rolling high plains of eastern Wyoming. I asked a local cowboy, boots, stetson, and all, the directions to Exxon's Highland Mine.

He studied me awhile with narrow blue eyes half hidden beneath the wide brim and said slowly, "Why? You gonna be workin' out there?"

"Might be," I replied, getting the distinct idea that cowboys and miners might not get along. "Why do you ask?"

"Just wonderin'. Seems to be lot of you out here working in them mines." Then he adjusted his hat, and said in a somewhat more amiable voice, "Just take the Orpha Road. Only road goin' out that way."

"How far?"

"Not far. Maybe thirty miles. There's signs out there, just follow 'em."

"Thanks, Pardner," I said. Thirty miles was exactly what I was afraid of. Why the hell couldn't the pioneers have built a town that was going to survive near a mineral deposit?

The drive to Exxon's Highland Mine covered a lot of prairie that looked virtually the same as it must have looked a century ago. The gently undulating plains, covered with a ragged carpet of coarse, gray sagebrush, were broken only by long, straight barbed wire fence rows and an occasional lonely windmill. The infrequent ranch houses, along with their surrounding clusters of sheds and barns, were huddled low along the creek beds out of the Wyoming wind. Most of the fence gates that faced the paved road, each marking twin ruts of tightly packed dirt leading into the sprawling private ranches, were posted with not uncertain signs about private land, no hunting, no trespassing, and keep the hell out. Although the short turnoff to the mine was paved, the asphalt on the main road ceased abruptly, and I noted that the next honest-to-God town beyond that point was Gillette, fully one hundred miles farther. I imagined the land five miles out of Gillette to look exactly the same as it did here. A band of Sioux dragging travois in their eternal search for buffalo would not have looked out of place.

The Exxon signs led to a few square miles that had been bladed with dirt roads radiating from a complex of aluminum prefabricated buildings, their gleaming metal surfaces totally incongruous with the drab gentleness of the prairie. The personnel building was clinically clean, lit with fluorescents, painted in soft pastels, and identical to personnel offices for large corporations in downtown anywhere. I received a warm sincere smile from the receptionist, who took my name. She was obviously a professional receptionist and not just some local cowgirl they told to sit at the front desk and grin when somebody walked through the front door. This was a total departure from any other mine personnel office in my experience, and I began to think that just possibly Exxon, in its multi-billion-dollar blue-chip splendor, might manage a mine that would exceed my dismal expectations. I made myself comfortable on a couch near a stack of mining and uranium publications, and exchanged a few professional smiles with the professional receptionist.

"Mistah Voynick?"

I looked up at the source of the deep southern drawl. He gestured me into his office. After the formalities of introduction I took a seat while he spread a file on his desk. I recognized the forms I had filled out earlier. That file, already quite extensive, was mine. After taking a second to glance over the papers, he leaned back in his chair and gave me a look of appraisal.

"Well, Mistah Voynick, we've had a little tahm to check out yo' previous employment, uh, Climax and Hecla, and to confirm the information in yo application. Now then, Ah see you've worked as a minah fo', let's see . . . "

"About two years," I interjected.

"Well, not quite that, more like a yeah and a half."

"More or less."

"Ah also see from yo' previous employment listings, uh, at least from what you put down, that you, uh, don't seem to spend too much tahm in one place," glancing upward to check my reaction.

"Well, I've been doing some traveling for the past several years, and some of those jobs just involved traveling."

"Tell me, now, this job you got down heah in the Caribbean, uh, marine salvage. What's that have to do with mining?"

I went into a long disertation describing the remarkable similarities between marine salvage in the Caribbean and underground uranium mining in Wyoming. When I finished, I was surprised myself to learn just how closely related the two fields actually were. I was also beginning to realize that this guy, in the name of Exxon, had a distinctly different approach to the hiring of underground miners. Anyplace else, upon confirming that I had two functional arms and two functional legs, would have put me on the payroll, given me some gear, and showed me the way to the cage.

The personnel manager, excuse me, director, was silent for a minute, possibly mulling over in his mind the amazing and little-recognized similarities between Caribbean marine salvage and Wyoming mining. "Well, now, Mistah Voynick, keeping in mind all this traveling you've done, just why is it you want to work for Exxon in Wyoming?"

Because I need five grand and you had an ad in the paper, I thought. But what I said was, "Well, I've got a lot of friends here in Wyoming and I'm thinking of settling here. And I've talked to some people who worked for Exxon, not here, but different places in the country, and they all said that the company benefits were about the best going."

That rang a bell. "That's very true, about the benefits, very very true. It's also a point we'ah all proud of heah at Exxon," he said with obvious pride and sincerity in his voice. "You see, Mistah Voynick, what we'ah trying to do heah is accumulate a nucleus of good people, people who are heah for the long term, people who can recognize a good situation when they see one. What we don't want are people who will be heah today and gone tomorrow." He said that last sentence with his eyes locked on mine to emphasize some apparent misgivings he had about my geographical instability.

We stared at each other for a few seconds. I was going to really pour it on, but decided against it. He spoke first anyway.

"Now then, Ah've got you' test scores and find them to be quite satisfactory, so we've got no problem theah. What I'd like you to do is drive ovah to the underground and talk to Wiley Brooks. He'd like to ask you a few questions."

"Fine, just tell me how to get ovah theah." He already had me talking in that damned drawl.

I was given a written pass and a white Exxon hard hat with "Visitor" stenciled across the front. The underground mine was about a mile or two from the open pit and mill facilities, which processed the ore to a concentrate right there in the middle of the trackless prairie. At the underground mine dry, I found an office littered with layouts and blueprints, mine lamps and yellow wet suits, and an inch of muck on the floor. The man who greeted me was tall, thin, and had a pair of alert eyes set in a weathered face. His handshake, as I expected it would be, was firm and sincere. He had that look of a man who had put in his share of time underground and knew what it was all about. I liked him immediately.

"Sit down," he said, clearing a wet suit top off a chair with a sweep of his hand. "Don't mind the mess, you've been in mines before. Gave up trying to keep this place clean. Haven't got time for everything." We talked for fifteen minutes, with Brooks steering the conversation expertly to find out what I had done and what I had not done in the underground. He pointed out that Exxon was just taking the mine over from the Harrison-Western development crews who had sunk the shaft down to six hundred feet and had gotten a station started. "We've got about a mile and a half of drifts down there now and we're gearing up to take the whole thing over ourselves. It's a little wet down there, but not enough to cause trouble. It'll be a lot better when we get our drainage systems and pump stations set up. And when we get done I expect we're going to have a good mine here, a good place to work. And we're going to have a safe mine, too. Now, we've already had a man killed here, as you may have heard, but that was over at the pit wall and had nothing to do with the underground. This is a safe mine so far, and that's the way we're going to keep it."

Wiley Brooks asked no questions about my stated non-mining pursuits or whether I had any plans for long-term employment. He had been around the western hardrock mines long enough to know the ways of miners. Probably some of the best miners he had ever worked with were "heah today and gone tomorrow." I returned to the personnel office while Brooks apparently telephoned his approval. Seated in the personnel office again, I listened to the drawl, which, this time, was backed with enthusiasm.

"Well, Mistah Voynick, Ah'm glad to see theah was no problem ovah at the underground."

I told him I had been impressed with Brooks as a supervisor and that, from what he had said, it would be quite a mine when it was developed.

"Ah've been underground mahself, and Ah know it's a good mine," he confirmed. I wondered how eighteen years with Exxon in the administration of oil fields and refineries could possibly prepare one to differentiate between "good" mines and "bad" mines. Wisely, I declined to inquire.

"Well, Mistah Voynick, Ah think we've covered everything and Ah am now prepared to offer you a job as an underground miner with Exxon. How does that sound?"

I said it sounded fine and asked when I would start.

"Now then, what you'll have to do is send us a formal lettah of acceptance. When we have that, and aftah we've been able to review the results of yo' physical examination, we'll be able to arrange a starting date."

On the drive back to Colorado, I couldn't help but think how different the entire Exxon way of life seemed from the other mines I had had experience with. As the personnel manager proudly said, this was Exxon's first effort in direct management and operation of a mine, and they were going to impart some innovative approaches to mine personnel management and reduce the absurd rate of turnover among underground workers. It would be nice, I thought, if they could succeed in imparting some innovative ideas to the underground workings while they were in the process.

The elaborate physical examination required by Exxon took two days to complete, and after the report had been submitted, along with my formal letter of acceptance, I sat back and waited.

"Yeah," I told one of my old Leadville friends, "going back in the mines for a little while."

"Where?"

"Exxon. Up in Wyoming."

"Exxon? The oil company?"

"That's the one. Everybody's favorite oil company."

"They into minin', too?"

"Just started. Uranium."

"You gotta be crazy to work in a uranium mine."

"No crazier than I'd have to be to work in any other hardrock mine."

"Yeah, but that's just it," he explained, "uranium ain't hardrock. It's all that ratty sandstone. Bad ground, comes down all the time. Good place to get hurt."

I went into a brief description on the hiring and how I thought Exxon might be able to run a better mine.

The old man snickered and said, "You know, you're even dumber than you look. What the hell is Exxon going to have for you? Easy chairs for you to sit your lazy ass in down there? A bunch of naked broads fannin' fresh air in your face? It'll be a mine, Pard, like any other goddamn mine, except maybe a little worse. And remember you're gonna be breathin' all that crap, too."

"Breathin' what crap?" I asked.

"That radioactivity crap."

"They're supposed to have all that controlled now," I answered, both for his sake and mine.

"What the hell do you think they told all them boys back in the fifties? The same thing, don't worry about it. That was back when everybody was in a big rush to make a million dollars and all them little uranium mines was goin' full blast in west Colorado and Utah. All them boys got for their trouble, beside a lot of rock on their heads, was lung cancer. Half of 'em anyway."

I sat back and digested the words of wisdom from the old miner, deciding that there might be a fair element of truth in what he had said. Any enthusiasm I may have had for beginning work at Exxon's Highland Mine was rapidly diminishing.

A week later, on an October Monday morning at 8:00, I reported to the personnel office at the Highland Mine once again, this time for two days of orientation before going underground. An unusual pile of forms had to be filled out to secure the benefits that came with the job. When we were going through the routine, the man across the table, who had also just hired on, asked me where I had come from.

"Worked down in Arizona a couple of years ago."

"Yeah? I'm from Arizona. Where did you work?"

"Hecla Lakeshore."

He smiled at the name, then said, "I almost went down there myself once, heard a lot about the place. How did you like it?"

"I didn't."

"You there for that fire when them two hands got it?"

I told him I had been, and we got into a long conversation about Arizona. Stan Griffith had worked at Magma Copper, and, although he was a few years younger than I, had put in nearly eight years in the mines. With the depressed copper prices, the Arizona mines had slowed down and Stan decided that Wyoming, in the middle of a uranium boom, would be the place to look for a decent contract or maybe a job as a shifter. It wasn't the first time Stan had tramped. He had a wife and five kids, and I had to admire him. Our conversation was interrupted by the introduction of the company safety man, who would give us the standard spiel. He was introduced in hushed, almost reverent tones as a man who had received countless awards for his accomplishments and efforts in the mine safety field. Exxon, and we as new miners, were fortunate to have him at the Highland Mine. He

Miners crowd onto a cage at the beginning of a shift for the ride down. Eight hours later there will be smiles on those faces. (ASARCO Incorporated)

gave a notably lackluster and unimaginative hour-long talk on the company's utter preoccupation with safety, their sterling record to date, and the use of self-rescuers. He also gave a general rundown of the local safety regulations. When that was done, we were told to pick up our gear and show up tomorrow morning ready to go to work underground.

Whatever ideas I had of this mine being anything different from the others faded away at the man cage, the small cage that ran adjacent to the large material cage. The man cage held seven miners packed like sardines. The bell signals were given by extending your arm through a hole in the protective steel mesh, jerking a code on the bell rope, and quickly withdrawing the vulnerable arm before the stomach-wrenching descent began. Each hoistman had a distinctly different touch, sort of a signature, on the levers that released the huge cable drums controlling the vertical movement of the cages. Whoever this one was, he had the touch of a plumber. Every miner on the cage, as if on cue, simultaneously winced and sealed his lips as if to keep the contents of his stomach from oozing out the corners of his mouth. The only difference I could see between the Highland Mine and the others as the cage rattled and banged along its tracks in the pitch-dark, drafty shaft was the tiny Exxon stickers everyone had on the grimy hard hats. Exxon or not, the smell of that shaft, the characteristic heavy, musty, damp odor that permeates everything, told me that this one was going to be no different at all.

Gerry Miller was my shift boss, and he promptly put me in the incline heading with two other miners. Gerry struck me as somewhat of an incongruity down there, always neatly dressed, clean shaven, and soft spoken. He was the kind of guy who would be better cast as a supermarket manager, someone whose honest smile and sincere concern were more appropriate for calming irate housewives than bossing a mine crew. The first thing I wanted to ask him was, "What's a nice guy like you doing in a hole like this?"

The Highland Mine was still in the ground floor of its development, with about a mile and a half of drifts winding out from the shaft station. There were only two active headings worked by Exxon crews. A third was being finished up by a Harrison-Western crew that was getting ready to pull out and turn the whole show over to Exxon management. The incline heading, which was going to be home for me, was a twelve-foot-square drift running uphill at a medium grade through a coarse green sandstone. The drilling was done by jackleg, and the mucking and hauling of supplies were done with a load-haul-dump unit called a 911, a diesel-powered machine similar to a severely undernourished ST-8. Everyone at Exxon seemed impressed with the 911. Before I ever saw one, I was warned about their size and power.

"They're pretty big machines so be careful with 'em, Pard," I was told. When I did see one of the loaders, I couldn't help but smile. An ST-8 would eat two of them for breakfast and look for more.

The first couple of shifts were spent playing around with equipment that didn't work, advancing service lines, installing vent bags, playing with plugged pumps, and maybe getting a round out every second shift. There was no big push; everyone bided his time and seemed to be waiting for something called an Alpine Miner to arrive on the scene. Meanwhile, the incline advanced with the conventional system of drilling and blasting.

"When we get the Alpine in here, it'll be a little different," Gerry told me.

After my curiosity got the best of me, I asked what in the hell this eagerly awaited Alpine Miner was. For all I knew, it could've been a little German guy in short pants holding a beer stein and yodeling.

"Don't you know what an Alpine is?" he asked incredulously.

Accepting the disgrace, I told him I didn't have the faintest idea.

"Did you notice the yellow machine by the shaft station with all the crap piled on top of it?"

"I thought that was some junk waiting to be hauled out."

"Oh, no, that's it. The Alpine Miner. The mechanics are waiting for more parts. When they get it together it's going to make a big difference. We'll be getting out a round and a half each shift."

On the way up to the surface for lunch, I stopped at the station to take a good look at the Alpine Miner that was going to make everything easy for us. For the life of me, I couldn't imagine that wreck's doing anything. If Gerry thinks so, it may be. In the meanwhile, it was jacklegs and dynamite.

Lunches for all underground workers in uranium mines were required by law to be eaten on the surface, which provided a nice midshift break. The crew was small, not more than twelve miners, and since there was no lunchroom, the benches in the dry sufficed. Everyone would gather in one corner to spread out his lunch, pour coffee, and tell lies. The dry, like all mine change rooms, had the pungent smell of a football locker room after a rough game. It had a rich, humid, fetid redolence that spoke of dead socks and jocks, pools of sweat, and abused long underwear that, if given the chance, could probably drive a heading by itself. It was hardly a noon luncheon, but rather a functional ingestion of nourishment to provide the strength to last the final four hours of the shift. Someone told me a man could get all the nourishment he needed just by breathing the air in the dry.

Eating and smoking underground were forbidden in all uranium mines by the Bureau of Mines, as a precaution against the radon hazard. The Highland Mine had not yet entered the ore body, and if there was a measurable radon level at all, it was negligible, but the regulation was mandatory anyway. All the ores of uranium and the ores of other associated heavy radioactive metals, through the natural process of radioactive disintegration, produce radon gas and its isotopes, known as "radon daughters." These gasses are tasteless and ordorless and may be detected and measured only with special equipment. Radon, however unobtrusive, is not at all harmless. The inhalation of sufficient concentrations has been proved to induce lung cancer.

The radon problem in the early uranium mines was a space-age parallel to the silicosis scourge that wrecked the lungs of the nineteenth-century miners. After World War II, the sudden demands for uranium ore, which could be processed into components for bombs and for the embryonic reactor, touched off one of the last mineral rushes in the U.S. In 1950, the picture of the modern prospector was of a man prowling the remote, dry mesa country of western Colorado and eastern Utah, holding a Geiger counter and waiting for the telltale click-click-click that would make him an instant millionaire. It was a good bit of romance that captured people's imagination, but it ended right there. What happened in the hundreds of small underground mines that were rapidly developed to recover the ore never came to the public's attention. Even today, a complete knowledge of radioactivity has still not been achieved, but back in the 1950s such knowledge was elemental at best. As a result, the miners were faced not only with the well-understood spectre of falling rock, which, in the ratty sedimentary deposits that contained the ore, was a common occurrence, but also with the largely unknown dangers of radon gas inhalation. As far as the mining companies went, the idea was to get out the most ore in the shortest time at the least expense. What happened, unfortunately, was a modern "silicosis," which afflicted the miners in the form of a very high rate of lung cancer. Many of those early uranium miners are today making it on one lung, and the unlucky ones—or lucky, depending how you look at it—collected the big ticket to premature retirement.

Since it was impossible to prevent the miners' exposure to radon gas, and since the ore was going to keep coming out, exposure limits were set and radon levels in the underground environment were regulated and measured. The control of the radon levels was then, and still is, managed simply through increased air ventilation, which serves to dilute the concentration in the air the miners breathe. Various techniques to suppress the seepage of radon from the rock have been tried.

They included pressurizing the mine atmosphere and spraying sealants on the rock surfaces, but both were finally deemed by the company to be impractical, ineffective, or prohibitively expensive. Today, the ventilation-dilution method is still relied upon to reduce radon concentrations to what is believed are acceptable levels. As the radon atoms were shown to have a tendency to adhere to airborne particles and to surfaces in general, smoking and eating in the underground were forbidden.

Uranium mining itself is not hardrock mining in its purest sense, although most of the miners who make up the uranium circuit have had hardrock experience, and there is a free exchange of men between the two kinds of mining. Most of the techniques used in underground uranium mining are also from the hardrock school. The uranium ore bodies themselves were formed through a complex geological phenomenon which began with the formation of the Rockies. In the simplest terms, as the mountains wore down into alluvial flows, the uranium mineral compounds tended to concentrate, because of their great weight, similar to the way gold placer deposits form. Since the uranium compounds are quite soluble in water, leaching further concentrated the ores into the present deposits that are economically feasible to mine today. Most underground uranium mines in the western U.S. are relatively shallow, with few shafts exceeding one thousand feet in depth. The drifts are driven out through sedimentary strata, what we commonly refer to as sandstone, beneath the ore body. A cave system of mining is used to extract the ore. On paper, it is all very clean and very simple.

From the day I started at Exxon, there had been a lot of talk about the contracting that was going on at Kerr-McGee's Bill Smith Mine, only eight miles away across the sagebrush. Any effort to get more work out of the Exxon miners was usually met with, "I'll start breaking my ass the day this outfit comes up with a contract." Exxon had made up its mind that the Highland Mine would be run on a day's pay basis, apparently believing that its long-term benefit program would be sufficient to keep miners from demanding a contract. Exxon, with the best of intentions, had organized the mine and conducted its recruiting effort along the lines it had developed through decades of experience in managing oil fields and refineries. This was their first encounter with the peculiar breed of man known as the hardrock miner. The basic concept of Exxon's employment policies was to foster a long-term company loyalty, with the carrot being the mountain of benefits one would slowly accrue over the years. The idea worked all over the Gulf Coast and the East Coast, so why not carry it through to the underground miners who would be hiring out at

Highland? But there is little if any similarity between refinery workers and hardrock miners.

Because of the basic environment of an underground mine, and because of the everyday dangers that a miner looks in the eye, most have little reason to think in terms of what they will get ten or twenty years in the future. There is a distinct possibility that he won't be around, or at least not in one piece if he is, to collect and enjoy all his company bennies. Historically, miners have always lived for today; tomorrow would take care of itself. Since a miner takes his daily risks and trades a bit of his health most every shift, it is only natural that he should want all the money he can get. A white- or blue-collar worker can sit for twenty years and watch his retirement creep closer. He knows that, unless he gets hit by lightning or chokes on a ham sandwich, he'll be around for the gold watch and the pension. It is easy to take the "rough" times in the office or the shop with a grain of salt and keep counting the days. On the other hand, a miner, after a rough shift in rotten conditions, or maybe after watching his partner get crushed or buried, can have very little company loyalty. The dream of retiring on some bright day in the hazy future can seem like an utter absurdity. The miners are working for that next check, and if the mine down the road is willing to come up with a better check, there is not much of a decision to be made. All underground mines have essentially the same conditions and dangers. Some may be hotter, colder, wetter, or drier, but they are all from the same mold. The only thing that really matters is the check on Friday.

Ever since the mining circuit was established on the western frontier, the miners who have worked it have always been wanderers. The many boom towns that sprung up overnight clamored for experienced miners, but few of these towns enjoyed any longevity to speak of. They were boom towns one day and ghost towns the next when the ore ran out. The miners took all this in stride. They packed up wife and kids and whatever they owned and tramped over to the next district or to whichever camp was making the most noise. The miners on the Exxon crews represented the Homestake in South Dakota, the Coeur d'Alene district in Idaho, the Arizona copper holes, the Colorado molybdenum mines, and other uranium mines in Utah and New Mexico. Tramping to Wyoming meant nothing, and tramping out of Wyoming would mean nothing. I never met the miner who considered such moves a major relocation. It was simply a part of the game. Exxon's employment policies and goals were noble indeed, and doubtlessly would have met with instant success if the guys in the hard hats and lamps had been refinery workers. The policies included, initially, a no-rehire clause for miners who tramped out. In comparison, Climax, which was as organized as a mine can get, put a limit of three tramp-ins and tramp-outs on a miner. In the bull sessions in the dry, it was generally agreed that Exxon would have a rough time trying to make oil workers out of its hardrock miners.

Work went slowly in the incline heading while we all waited for the mechanics to effect the resurrection of the Alpine Miner, which still rested at the shaft station with junk piled all over it. One of my partners in the heading was Bill Cook, a slightly built man with a perpetual faint grin you had to look for. Bill, who had gotten most of his experience in the uranium mines at Grants, was a nice guy once you got to know him. That wasn't the easiest thing to do, since he rarely wasted words. He was one of the best miners I ever met.

Although he was an excellent all-around miner, Bill's forte was the jackleg. To appreciate a man's skill and competence on the jackleg, one must first understand a little about the machine. A jackleg has always been to the underground miner what a wrench is to a plumber or a scalpel is to a surgeon; one's skill with each is the most visible measurement of practical ability in each field. Properly called a feedleg drill, the unit has taken on the popular name of "jackleg," the name given to the machine by one of the major manufacturers. Basically, the jackleg is a pneumatic rock drill that runs on one hundred pounds of air pressure. It will accept drill steels from the eighteen-inch starters to the lengthy ten-footers. The drill is mounted with a swivel joint on a pneumatic leg, which, when compressed, is about four feet long. The leg, when given air, will extend telescopically to a maximum of about ten feet, and is used to maintain pressure on the drill steel as the holes deepen in the rock. The entire unit, drill and leg, weighs about 130 pounds without the steel and the trailing weight of the water and air hoses. Operated by a person skilled in its use, the jackleg can punch holes 1½ inches wide in rock that would have frustrated the old-time miners who worked before its introduction.

The operation of a jackleg is quite simple, since there are only two controls to worry about: a lever supplying air to the drill, and a knob supplying air to the leg. But this simplicity of operation does not necessarily mean that a new hand, or even a miner with limited experience, can simply pick up a jackleg and drill out a face. On the contrary, the jackleg in the wrong hands can be a dangerous piece of equipment. Keep in mind the 130 pounds balanced precariously on a single leg, the clumsy trailing hoses, the loose muck piles, the sole source of light being your cap lamp, a vicious vibration, and a deafening roar that would blow the eardrums out of the four stone faces. I can recall with great clarity my first encounter with a jackleg back at Climax. Puffing in the rarified air, I dragged a jackleg up to the muck pile

A miner with a jackleg drill, a common sight in any mine. (Gardner-Denver Co.)

to drill a couple of holes that were needed, hoping to surprise my partner when he returned. In the month I had been underground I had watched the other miners running the machines. It looked pretty damned easy. Well, it wasn't. Before I knew what happened, I got knocked on my ass. Not about to take that from any drill, I picked it up and gave it another try. This time I got a death grip on it, turned the drill on full, then gave the leg some air. The air control for the leg must be manipulated very, very carefully, as only a slight difference in the valve opening makes a big difference in what the leg does. With the brain-rattling roar of the drill, it was difficult to be too subtle on the leg control, and I gave it too much. The drill took off like a goosed kangaroo with me wrapped around its neck. It was a brief ride, but a memorable one, ending abruptly when I smacked into the overhead cap. That jarred me loose from the drill and I fell back, crashing into the muck pile with the drill right behind me, missing me by a foot or two. By that time I was getting pretty sore, and I decided against pushing my already overtaxed luck on another try. I would yield to prudence and await suitable instruction. That introductory meeting with the jackleg, probably because I didn't break my neck, now strikes me funny. But jacklegs are not always so tolerant of errors. At Climax a man placed his finger on the chuck of a drill and the machine responded by promptly removing the errant digit. On another occasion, the leg of the drill slipped from its base in the muck and, instead of propelling the drill forward, shot out to the rear, nearly penetrating a man's chest with the steel forks. He was quite lucky in receiving only two broken ribs and one of the nastiest bruises I have ever seen.

With enough practice, almost any miner can drill out a face with the jackleg. The difference is in the work the miner does himself, whether he wears himself out horsing the unwieldy drill around, or whether he uses the inherent mechanical capabilities to their fullest. There are damn few miners, regardless of what lies they tell, who are true masters on the jackleg. One was Kenny Reasoner, who, back at Climax, put on a memorable demonstration under a notoriously bad back. Another was Bill Cook.

"Well, Pard," he'd say at the start of a shift, "we got to find us a machine." He'd start looking up and down the ribs, in different headings, and in the crap that was always piled up at the station. Bill wasn't looking for just any drill, because not just any drill would do. "No, not that one, Pard," he'd say when I dragged one into the heading, "something's wrong with the leg on that one." How he could tell the difference at a glance was a mystery to me, since all the drills were just amorphous masses of muck and grease anyway.

When Bill found *his* drill, a look would come into his eyes as if he were greeting an old friend. As small as Bill was, he would easily swing the drill onto his shoulder in perfect balance and haul it back to the heading. Then he would hook up the water and air hoses, sort out his steels and bits, and get ready for something he was obviously going to enjoy. "We got any safety chains around here?" he'd ask.

I'd look around, knowing it was a futile search. According to Exxon's safety regulations, safety chains were required for both the header end and the drill end of all air hoses. They were particularly necessary on the drills, where the vibration was capable of loosening the butterfly nut no matter how well it was tightened. If an air hose did work loose, it would slash around the drift like a demented snake, driven wildly by the screaming force of one hundred pounds of air. Miners would hit the deck and cover their heads, trying to crawl as fast as they could, while avoiding the whipping hose, to reach an air valve along the rib. If the two pounds of steel on the end of the hose happened to catch a man in the chops, it was guaranteed to scatter a mouthful of pearls on the drift floor. Management was always harping about the safety chains, but never seemed to supply any. I would like to have seen what would have happened if a miner refused to drill because there were none. He would likely have been viewed as a lazy son of a bitch trying to get out of some work.

"The hell with it," Bill would say when I came back empty-handed. "I'll just keep an eye on it." Moving the drill up to his pre-determined spot, he'd tell me, "Watch it, Pard, get back out of the way a little." He didn't want me out of the way so much for any safety consideration, but rather so that I wouldn't cramp his style.

The pounding roar would fill the heading and echo down the drift, the air at the face would become eerily misted from the oil and water fog spewing out of the drill, and Bill Cook would be in his glory. Bill never wore ear plugs or protectors, as few really good miners did, but preferred to catch every sound, every slight departure from the norm, every little deviation that might indicate something was wrong either in the drill, the leg pressure, or the rock itself. Every once in a while, a five- or ten-pound chunk of rock would jar loose from high on the face and flash through the mist in front of the drill. Bill would rest his gloved hand on the rear handle of the drill, not forcing, but just sensing the vibration of the thundering machine. He would adjust the leg, a little more or a little less air, ever so slightly, measuring its effect by the almost-imperceptible change in the tone of the loud roar. When everything was right, the way he wanted it, the way it should be, when every bit of what was engineered into that drill was coming out of it and taking its toll on the rock, only then would Bill Cook turn to me and smile. It was a smile of utter contentment and satisfaction, the kind of smile that would steal over the face of a symphony conductor when all 101 instruments were performing as one.

The sandstone was relatively soft compared to hard rock, and too much leg pressure could easily cause the steel to plug and hang up. Bill would tread the narrow limit of leg push, giving it all it would take and never plugging a steel. On completing a hole, he would never accept help in pulling the heavy machine back out to start a new one. He would always manage by himself with a minimum of visible physical effort, a totally coordinated concert of gravity, balance, angle, and pneumatic power. I swear those drills knew when Bill Cook was running them. Like the expert horseman who becomes a part of his mount, Bill became a part of his drill. The combined fluidity of motion and grace made the operation of rock drilling seem like an effortless dance.

Other miners, including some of the biggest, burliest ones with years of experience, were good with the jacklegs. But in watching them one noted the physical straining, the tiring lifting of dead weight that Bill never seemed to do. Every other miner, at some point in the drilling operation, would wind up applying to the drill a combination of every sexually related adjective known to man. Bill would never talk to his machine like that. Rather on the rare occasions when something was amiss, he would examine it with the concern a father might give a child, determine the problem, tenderly correct it, and when everything was right again, turn to me and smile. Watching Bill Cook drill was highly entertaining. Those were probably the only times I could be even remotely enthused about doing it myself. But when I did, invariably the jackleg was always the same greasy, unruly, 130-pound monster just waiting for an opportunity to knock me on my back again. The best I'd ever do was fight it to a draw.

Blasting was the other operation in which Bill took pride and pains to achieve perfection. Since he had drilled the pattern, all the holes were placed where they should be. Because of the water in the mine and the chance of stray current, rather than shoot electrically we would often time the round with regular fuse. Each primer would have a precisely measured length of fuse to insure the order of detonation. When the round was loaded, the face would be covered with red fuse strands all joining in the center where they would be ignited simultaneously. When they were all sputtering, smoking, and spitting red sparks, as many as thirty fuses at once, the face would look like a big Christmas tree. Bill would stand in front of the loaded face with this look of serenity on his face and love in his eyes, as though only he could hear the angels singing the *Ave Maria*. He knew in his heart that the rock would break the way he wanted it to; there was no other way. When the glittering, deadly red sparks had disappeared into each drill hole, he would turn to me with a look of great sadness. It distressed him that he would be unable to witness firsthand what he and the DuPont people had wrought. He would then motion us away. From the station, we would listen to each timed charge detonating. The entire round would take about twenty seconds to go, and Bill would listen to each separate detonation with a vacant, detached smile, occasionally giving me a glance that reflected supreme satisfaction when one sounded particularly good.

Bill loved the drilling, blasting, and timbering, but was definitely annoyed if required to perform some Mickey Mouse crap, which happened often. He was basically a contract miner, and he also expected that everyone should approach his levels of perfection, particularly the other shifts who worked in our heading. The "other" shifts, as any miner can tell you, are a bunch of asses who don't know the first thing about mining, and it would upset Bill no end to come in at the start of a shift and find the heading in bad shape. That is exactly what caused Exxon to lose the man who was probably the best miner they ever had.

"Will you look at this crap?" he asked me, standing dejectedly in front of a lopsided muck pile. "Will you look at that face?" The face was a jagged, irregular mass of cracked rock. The last shift had really butchered the blasting, and the face would require considerably more work before a regular round could be drilled out again. The job at Exxon was little more than a hobby for Bill, who probably had his mind on a shifter job or a contract somewhere. All he said was, "I can do without this," and then he walked out of the heading. An hour later someone asked me what happened to my partner.

"Nothing happened to him that I know about. Why?" I asked.

"I hear he just tramped out."

"Bill?"

"Yup. Mumbled somethin' about asses and said he'd rather wait for that United Nuke shaft to get started."

Later on I found out that Bill Cook had indeed quit. Management had pleaded with him to stay on, assuring him that everything would be better, but Bill would have no part of it. I doubt whether Exxon ever realized exactly what they lost when he quit. I never saw him again.

Finding a place to live within reasonable proximity of the mine was an interesting problem. The choices were limited to three towns: Douglas, Glenrock, and Casper. About half the crews lived in Douglas, with the remainder divided between Glenrock and Casper. All three were located on the North Platte River, in keeping with the historical pattern of the region. The valley of the North Platte was the avenue for the first fur trappers and mountain men who explored the Rockies. It later served as the route for the Oregon Trail and for the first railroad constructed through the region. The river carries a good volume of water all year and supports good growth of trees along its valley, making it an oasis of sorts, a pleasant respite from the expanse of barren, semi-dry prairie on either side, particularly the north. Aside from those three Platte River towns, there is no other place to live for a hundred miles.

Wyoming, as the highway welcome signs with their bucking horse and rider proudly announce, is the Cowboy State. Indeed, it is aptly named. And if Wyoming is the Cowboy State, then Douglas should be the Cowboy Town. Its three thousand residents cluster around one movie theater and twenty gas stations, which are in turn grouped around the five bars. An eight-foot-high jackalope, one of those jackrabbits that a taxidermist screwed a set of antelope horns into, sits in the dividing island of the main drag in the middle of town. Cast in cement, the jackalope keeps a wary eye on the cowboys, miners, and oil rig workers. Knowing full well the origin of this composite beast, I inquired about it anyway for the sake of conversation when I checked into one of the dive motels.

"Quite a rabbit you have out there," I commented to the little round-faced manager.

"Oh, you mean our jackalope?" was the reply. "You know what that is, don't you? That's a cross between a jackass and a cantaloupe." That was followed by peals of convulsive laughter. "Get it? Jackass and cantaloupe? Jackalope? Heh, heh," he explained rapidly as the laughing died out. When he saw that I was not about to join in an accolade of his wit, his moon-shaped face reddened and he cleared his throat. "Well, heh heh, that's what I tell a lot of the tourists. They get a kick out of it. Sometimes. Uh, gonna be staying with us or just passing through?"

He had made up my mind. "Just passing through," I said. That decision was quickly confirmed by a walk around town. After the next shift, I headed toward Casper, stopping at Glenrock along the way. Glenrock was less than a quarter the size of Douglas and had progressed little beyond its stage station beginnings. If a stage coach rolled down the middle of town, I doubt whether anyone would have looked twice.

Casper, with 41,000 people, was the biggest city in the state and, as far as I was concerned, was the only place to live. It was a good drive from the mine, about fifty miles, but the people here would at least look up and wonder about a stage coach. And they didn't have one of those damned jackalopes blocking the main street. Casper was bounded on the east and west by oil refineries, vestiges of the days when it was a booming oil capital. Immediately to the south were pine-clad mountains, and to the north were a couple of hundred miles of rolling prairie.

Wyoming is possibly the only western state that has maintained an honest cowboy image. Arizona and Colorado, for example, two other states that have had their roots entwined around the cowboy and frontier legends, have been softened and changed by the influx of tourists, skiers, retirees, sun seekers, and a stream of new arrivals seeking work in the industry that surrounds both Denver and Phoenix. Aside from the Jackson Hole-Yellowstone western extremity of Wyoming, the remainder of the state has undergone little change, and is exposed to a minimum of outside influence. There is still no manufacturing industry to speak of, and most of the jobs that are available are of the heavy sweat or outdoor variety. Wyoming was founded on the open-range cattle industry, which gave the state its cowboy base. After the turn of the century, the oil fields, some of the richest and most extensive in the country, were brought in within fifty miles of Casper, drawing an avalanche of rowdy oil workers from California and Oklahoma. The next industry to open up in Wyoming that drew a large number of new residents was mining, not the early gold strikes in South Pass and Atlantic City, but large-scale modern mining. The underground trona mines have completely changed the face of Rock Springs and Green River in the southwest corner of the state, the huge open-pit coal operations have brought life to Hanna, Gillette, and other towns, and the uranium mines are scattered through the state at Jeffrey City, Gas Hills, Shirley Basin, and the Powder River Basin. Many Wyoming towns are composed primarily of a colorful combination of cowboys, roughnecks and miners, all of whom are essentially "cowboys" in the general, romantic sense of the word, men who come and go as the jobs come and go and who are concerned with little beyond the paycheck on Friday. They have a basic untempered earthiness, a quality which is dwindling in the West, and a contentment with the simpler things in life, those diversions which offer immediate rewards, tangible or intangible. In Wyoming these are taken to be women, guns, trucks, grass and booze, the order of preference varying with the individual. Any lack of sophistication and pretense is replaced by a straightforward honesty and candor. These people have neither the time nor the predilection for phoniness or facades. If they like you, they will tell you so, and if they don't, they will also tell you so.

Wyoming is booming, and every day more and more people are moving into the state to take the jobs on the drill rigs or in the mines. The large number of transients has unfortunately resulted in an abnormally high violence and crime rate, which the state is trying its best to cope with. Housing for all these people is also a serious problem and the mobile home dealers are cleaning up. I was very lucky to find a good apartment within a week.

The drive to the mine was about fifty miles, but could be done in very little time, as the Wyoming roads encourage flying low. The car pool would take Highway I-25 to Glenrock, then cross the Platte and head northeast across the prairie at eighty miles per hour past the windmills, groups of delicately colored antelope, and the mobile drill rigs that would move on to another location every few days. Core and exploratory drilling were being conducted at a breakneck pace. The energy shortage and resulting boom had acted as a catalyst, it seemed, to nearly everything in Wyoming, the oil and gas, the coal and uranium. We often joked about the lot of the poor ranchers, not having enough room to run their cattle between the drill rigs, oil wells, mine headframes and coal pits.

On the ride to the mine, half of the car pool would sleep and the other half would stare morosely out at the prairie, with its rolling sage dotted by antelope and windmills, and wonder why. The morning radio out of Casper frequently played one song that got a rise out of most everyone. It was about cowboys, but everyone, in his mind, changed that to "miners." The idea was the same. When the slow, heavy guitar downbeat started, someone would always point it out. "Hey, here it is, listen to this."

> Mamas, don't let your babies grow up to be cowboys
> Don't let 'em pick guitars and drive them old trucks
> Make 'em be doctors, lawyers, or such . . .
> . . . Budweiser buckles and soft faded Levis
> And each night begins a new day
> If you can't understand him and he don't die young
> He'll probably just ride away . . .

"Tell me he ain't talkin' to us."

"He's talkin' to us, alright."

"Well, what's your excuse? Why ain't you a doctor, lawyer, or such?"

"Sonuvabitch if I know."

"I'll tell you, I wouldn't have to be a doctor or lawyer. I'd settle for anything that would keep me out of these goddamn holes and still let me eat right."

"Yeah, and pay for that car and fancy-ass camper you're buyin', too."

Silence again, only the hum of the tires and the sound of the Wyoming wind whistling through the windows, as everyone ponders just how it suddenly came to pass one morning that he had on a hard hat and a lamp and was running for the cage, how fate never really gave him a chance. The quiet will become uncomfortable and some-one will chance the thought.

"You know your partner ain't comin' in today."

"Yeah, he said something about that, goin' to court or something."

"I never got that story straight. Was that the game law violation?"

"Yeah, they caught him draggin' an elk through the corner of Teton Park."

"He shoot it there?"

"Swears he didn't. Says he shot it on National Forest land and was just draggin' it back across park land 'cause it was shorter."

"You believe him?"

A snicker. "Hell no, he probably shot it on park land."

"Damn right he did. You think it would stop him just 'cause that happened to be park land? Hell, the Denver Zoo wouldn't stop him."

Another car pool passes on the right. Faces framed in the frosty windows, eyes half shut, halfhearted waves and one finger. Real enthusiasm.

The incline slowly crawled upward through the green sandstone 680 feet beneath the sagebrush, the mine reverberating to the steady roar of the jacklegs and the concussions of the dynamite blasts. The Harrison-Western crew completed its contract work and the job of developing the mine fell fully on Exxon's shoulders. There was a standing offer to hire all the Harrison hands, but, interestingly enough, only one gave it a try. He lasted for two shifts and tramped out. The rest of the Harrison hands went right over to Kerr-McGee to hire on in a contract mine.

More and more, the lunch talk centered about what the Kerr-McGee hands made on their contract during the past week. United Nuclear was going to begin sinking a shaft in the area in a month or two; it was also going to be a contract job. With every shift we put in at day's pay at Exxon, there was more and more talk of tramping. It is one thing to work for a day's pay when the next contract mine is a couple of hundred miles away, and another thing to watch half the shift traffic turn off at the Kerr-McGee.

Jake Alcala, one of the miners who hired on with me and also lived in Casper, dropped out of our car pool and tramped over to

Kermac. Then Bob Hunter, from Douglas, packed his gear and did the same. But for all the hands that left, new ones showed up to take their places. Two of them, Larry Crabtree and Pete Hixson, were some of the first products of the Exxon School of Mines, as the company's underground trainee program was sarcasticly called by the miners. This Exxon innovation was the first underground mining school that any of the guys had ever heard of. The standard procedure for as long as the old timers could remember had always been to take a new hire and throw him in the hole. The education of a hardrock miner was as simple as that. If he didn't quit or get hurt or killed, the company had itself another miner. At many mines, brief safety orientations are given to familiarize a new hire with local regulations and procedures, but any instruction and learning takes place in the headings. Exxon's school was conducted on the surface at the wall of the open pit, and it covered such basics as drilling, blasting, and timbering, along with a lot of idealistic safety blurb, over a period of two months.

Whether it was a good idea or not was debated by all the miners. When a new hand comes underground for the first time, he usually has the hell scared out of him, and it is that reaction that fosters the attitude of extreme caution that tends to keep him alive. The pit wall school seemed to create a false sense of confidence among the "graduates," as well as a popular misconception that all of the safety policies were actually enforced. Pete Hixson was one of the graduates. He was a good-humored, albeit scatterbrained, twenty-two-year-old blond kid from Montana who had decided against another winter on an oil rig. After his two easy months, he was assigned to our mine crew. Seeing what the underground was really like was a bit of a shock.

"Jesus, I didn't think it'd be anything like this," he said in awe. "They told me all this safety bullshit for two months about how all these safety regulations were enforced and how no work was done unless it was safe. Stuff like that," he pointed to a half case of dynamite with a couple of detonators lying on top of it in a corner of the station that had been lying there for a shift or two, "well, they said you'd get fired for that. They told us that all of the regular miners took care of all this safety stuff first."

We were all amused by Pete's idea of what it was supposed to be like in the underground, and were really getting a kick out of his animated and honest answers to our questions. A diesel motor rolled by with two cases of dynamite on top, where regulations prohibit the hauling of supplies, and at the operator's side was a cardboard box of detonating caps. The operator gave us a nonchalant wave as the motor rumbled by. We were also prohibited from carrying both powder and detonating caps at the same time.

"See? Right there. They told us for a week that that stuff didn't happen and that if any regular miners caught us doing that they'd kick our asses up and down the drift."

"Well, stick around and you're gonna see things work a little different down here," one of my partners told him. "From that north heading, it's probably a half mile to the powder magazine and the cap box. Sure you ain't supposed to carry the two of 'em together, any clown knows that, but they expect that shot to go at the end of the shift. Take your sweet time and make two trips burnin' up an hour, they're gonna look at you cross-eyed."

Pete shrugged. "Yeah, well, I guess I didn't learn too much up there, but it was kinda fun. This old guy, I guess he used to work in mines all over, used to tell us bullshit stories about half the day. Everytime we thought he was gonna stop, we'd ask some kind of dumb question and he'd ramble on for another half hour. That safety guy used to talk to us a lot, too. He was the one who told us this safety program was really something."

Pete was there about two days when he almost picked up his early retirement. When his curiosity about what might lie in the blackness of the shaft got the best of him, he hung his head over the protective screening to shine his lamp into the sump. The shaft bells that are rung on the surface are repeated at the underground station for the express purpose of informing miners what the cage is doing, but that meant nothing to Pete. It was only the rush of air on the back of his neck that caused him to quickly withdraw his head a fraction of a second before the cage flashed by, smacking his hard hat. With his Adam's apple resting on the steel frame of the screening, he had missed decapitation by a matter of inches. When he told us about it later, he stood there with a foolish grin and said he had an Exedrin headache. Nobody said a thing.

Pete's mining career came to an abrupt end about six weeks later when he was helping manually load some 1,200-pound rails on top of an empty muck car. It is not likely he was told in his safety classes that we carried the rails loose on muck cars, either. The crew lost control of the rail they were lifting and Pete, who was on top of the car, had to get out of the way quickly. He did so by making a head-first dive into the steel bottom of the empty car and breaking his wrist. He took his six weeks of pay and never came back. The other new hire from the pit wall did something to his back very soon after going underground and his work attendance was sporadic at best. He tramped out a few weeks later.

It amazed me how proud of the safety department the people over in the personnel office were. I remember the introductory speech they gave about the safety man's competence and about the awards he had stacked up. During the four months I worked at Exxon, which means about five weeks of day shift when the safety department was also working, I didn't see that safety man underground more than three times. Whenever he would make one of his rare appearances, he would check the fire extinguishers, taking care to initial the inspection tags, take a short walk around avoiding the headings, then ride the cage back to the surface where there was nothing but clear blue Wyoming sky over his head. Most of his time was probably spent driving around the open pit making sure the Electra-Hauls weren't speeding, giving talks to the pit wall new hires, and filling out reports by the ream so he could get a few more awards.

A safety man, the head of a company safety department, has the authority to give a miner a couple of days off, or even fire him, for an infraction of safety rules. He also has the authority to inspect the underground and, upon finding something seriously below standards, say ventilation, to walk up to the super and tell him to shut the heading or the working down until he can get more vent rigged. Of course, that would be rocking the boat. Upper management is going to get upset when their tonnages and footages fall off, and they will ask not "why" but "why not." So the safety man can choose between two routes; he can be a safety man, or he can be a company man. It takes a rare individual to stand up to the company whose check he is collecting and to fully enforce the safety regulations despite the inevitable repercussions. Exxon had a safety program that was embodied in a magnificent set of files, but it was a program with no teeth. If they were content with such a program, fine, no miner was chained to the headframe, but listening to the pompous show they made of it could get awfully grating.

The near total lack of contact between the safety department and the underground was demonstrated at one of the complaint sessions that Exxon held every month. This gathering of miners, shifters, foremen, safety people, and management cost the company about a half hour of time-and-a-half per man. The meeting was just a general free discussion designed to loosen the lines of communication all around. The discussion got around to the brass system, where the miners picked up their brass at the start of the shift and turned it in at the end.

"Oh, no," the safety man said, "you've got that wrong. You turn your brass in at the start of the shift."

"No, we pick the brass up and take it underground," some voice corrected.

"No," the safety man said firmly to put an end to this temporary confusion. "We've always been turning it in at the start of the shift. I know. I set it up."

The awkward silence in the meeting room was finally broken by a foreman who was plainly embarrassed. "No, you're wrong," he corrected in a voice that was almost sympathetic. "We always pick up the brass and take it with us, like in every other mine."

More silence while the safety man's face turned a rosy red. "Oh," he mumbled at length, "I thought uh, thought that uh . . ." Everyone seated at the table exchanged glances and sly smiles. The unspoken thought was clear; if there was confusion like this on a basic point, what would the confusion be like if something serious were to happen in the hole?

As the weeks passed by, the snow began falling regularly and it was driven by a prairie wind that never seemed to let up. The wind would scream across the flatness of the high plains, driving the snow before it to stack up in huge drifts that would block roads in a matter of hours. Twice, the blizzards and ground drifts were so bad that Exxon called off two shifts at its own expense. The company did have a few benefits, and that was one of them. I always wound up on the right shift during the blizzards, getting paid for sitting in Casper and not for being snowed in on the prairie in the middle of nowhere. On one night before graveyard, our small crew showed up at the mine in winds that must have gusted to sixty miles per hour and higher. The electrical power was sporadic and, for a while, it seemed as if the headframe might literally blow over. We hung around the dry for four hours telling lies, nobody really wanting to go underground in those conditions.

As the incline crept monotonously forward, the routine would be broken every once in a while by something that would keep you on your toes. The rock had been getting progressively worse, and bigger slabs were coming straight out of the back. After a shot, the heading would usually require a lot of thorough barring down. On this occasion, a particularly large slab was hanging in the middle of the back, and I rode up in the bucket of the 911 to get it out. I was positioned under the protection of the last cap, but couldn't get the bar in for a good bite. Finally, I came up with a real brainstorm. Motioning the loader back a few feet, I stood up in the bucket and drew a bead on the slab from over the top of the cap. There are safety regulations which definitely advise against this practice, but I could see no other way to get that slab out of there without putting myself under it. Extending the six-foot bar over the cap, I drove it in the crack and began levering. Instantly, much more rock than I expected let go and crashed into the bar. The top of the timber cap acted as a fulcrum, and when the ton of rock drove the opposite end downward, the end in my hands flashed up with unbelievable speed, just missing my jaw but grazing my cheekbone and ripping my safety glasses, hardhat, and lamp off into the darkness. I sat in the bucket as the initial numbness wore off into searing pain. I was afraid to touch the side of my head for fear of finding it in several pieces.

"Hey, you okay?" my partner asked after he didn't hear anything for about ten seconds.

"Yeah."

After another bit of silence he again asked in a concerned voice, "What are you doing up there?"

"Just sitting up here in the dark holding my head, taking a break. Okay?"

"Yeah, fine. Long as you're okay. Damn, you sure got a way of gettin' that rock out of there."

"Glad you liked it." No blood, just some swelling starting. "Get me out of this goddamn thing."

He lowered the bucket and I crawled out, feeling around in the muck for my hard hat and glasses, my broken light dragging along behind me. As usual, it was a matter of inches that made the difference between a sore head and a broken head. I decided to go to the surface to replace my lamp and start lunch an hour early. At the station I had to put up with three guys who thought the swelling on the side of my head made me look like a 180-pound squirrel with a mouth full of nuts.

At long last the mechanics finished the job of resurrecting the Alpine Miner, and the electricians had hooked up the four-thousand-volt power cable that would trail the thing wherever it went. Gerry Miller walked up to me and asked, "How is my Alpine operator doing?" When I didn't say anything, he just smiled. "Don't worry, Pard, you're going to love it."

The Alpine Miner was a 28-foot long grinding machine with two tracks for movement and mounted a large, movable boom with two round cutting heads on the business end. The burr-like heads were about a yard in diameter and covered with two-inch-long cutting bits. Everything except the tracks and the conveyor belt was hydraulically powered, and the boom could be moved up, down, left and right while the cutting heads rotated. The operator could literally run the heads into the face of a heading and begin a cutting pattern that would rip and grind rock loose. The accumulating muck would fall near or on an apron, a large, flat plate mounted low at the front of the machine. Two crab-like steel arms would scoop it onto the conveyor belt, which ran down the middle of the unit. A 911 would wait with the bucket under the back of the belt to receive the muck and haul it off to a dump point. With not much practice, an operator could literally carve out the next round with just enough clearance to accept the next timber set. This wonderful machine would thus eliminate the need for drilling, blasting and a separate mucking operation. If everything worked right, a drift could be driven in a miraculously short time with nothing but the alternating steps of cutting out a round, timbering, and then cutting out the next round. The trouble was that nothing ever worked right.

The machine itself was not new. It had been through the wars and probably came out of some used Alpine Miner lot. The first problem was the shear pins that held the scoop arms and the conveyor together. The pins were designed to snap at the first stress from a rock jam to prevent any serious damage to the unit. During the first week of working with the Alpine, the shear pins would go every time we'd turn it on. On a good shift, we'd replace fifty or sixty of them.

"Got a few bugs to be worked out of 'er, that's all," Gerry would say as he walked away.

The tracks were another big problem. They would dig into the soft, wet sandstone floor, forming two ruts with a high center. Since the Alpine packed a lot of weight and rode low to the ground, it was only a matter of time until it hung up and became immobile. When that happened, as it often did, it would take four men, sweating and cursing, a 911, and an assortment of cables, jacks, timbers, and shovels to free the damned thing.

"If we can keep this drift floor a little cleaner, she's not going to get stuck every shift," Gerry would say as he walked out of the heading.

The final problem was the rock itself. As the incline moved upward we gradually ran out of the green sandstone and into some strata of lignite, a coal-like material, and rotten shale composed of almost nothing but fossils. The ground had very little support, and when it came down, it came down right, in six- and eight-foot pieces. Because of the bad back, it became necessary to shorten the sets from five feet to four feet, then finally to three feet. When the big slabs crashed down out of the back, the scoops didn't have a prayer of picking them up, and the only alternative was to manually break them up with a double jack. This was a clumsy and dangerous practice. The Alpine took up most of the drift and didn't give a man the room needed to swing the sixteen pound hammer. He was forced to stand very near the slabbing back.

"Don't take so big a bite with the heads and you won't bring that big crap down," Gerry would say.

— 194 —

It was unanimously agreed that the Alpine Miner was a monumental pain in the ass. The rock got worse, progress got slower, and the attention the incline heading received from management increased. On day shift it would look like a convention down there. Slabs the size of refrigerators came down, crashing on the boom to lurch the entire machine and throw pieces as far back as the controls. The more those slabs came down, the more I thought of Kermac.

The back had reached the point where it was just plain risky to timber the next set. Somebody once told me that a good miner would never have to get out from under the protection of his last set to erect his forward set. It makes good conversation in the dry, but when you're timbering you have to spend some time out there. The best you can do is to minimize that time. To make matters worse, since this was an incline and the timber sets leaned at the same angle as the face, a slab could come out of the back and hit three feet inside the last set. The only consolation in the incline was the fossils that fell out of the back by the ton. I would be fascinated by the endless procession of 600,000,000-year-old leaf and branch imprints that would grind by on the conveyor belt. I was disappointed that I could never find any fossilized animal remains, only plants. I certainly took the time to look for them and my partners thought I was nuts. During the breaks when they would lounge around in the rib telling lies, I would be breaking up shale looking for that one certain fossil that would set the world of archaeology and paleontology right on its ear. I never found it.

A few shifts later, Gerry must have decided we needed a demonstration in courage, and he came down to help us timber some very bad ground. At the start of the shift the back was talking regularly, and every few minutes a slab would let go and fall to the floor of the drift with a sickening thud. The hole in the back got bigger as the slabbing continued, and we suggested that if we waited a few hours until the first antelope fell through, there wouldn't be anything left to fall and it would be safe. Gerry didn't agree with that.

"Okay, let's go, we've got to catch that back now before it all comes in," he said, taking off his jacket and throwing it in the rib.

It is a simple matter to erect a set of timber in good ground. You can concentrate on the work without wearing your nerves out, and you can walk straight and not in a sprinter's starting stance keyed for the first sound or movement. Gerry and I stood on the cutting heads of the Alpine—I wonder what the safety department would have thought of that—and began cribbing off the cap we had just set. All of a sudden without warning, there it was, a rush of gray and a sensation of air movement on our faces as a couple of tons of rotten fossil shale let go. The rock smacked into the cutting heads and knocked us both off into the rib. When the dust cleared the two of us were limping around on battered ankles watching the swelling come up to fill our rubber boots. We hobbled up to the surface, cut the boots off, changed, and got into the company ambulance for the thirty-five mile ride to the Douglas hospital for X rays. On the way across the prairie we passed the sign marking the Kerr-McGee turnoff.

It's probably about time I looked into Kermac, I thought. Since we do all this driving through the snow to get out here, and since that rock in the incline is only going to get worse, and since we're on day's pay and Kermac is on contract, what's the sense in staying with Exxon? I looked for a reason, just one, and came up with a pension in twenty years. Twenty years in that decline, I thought, and I should be dead about ten times. When I got back to work in two days, I decided to talk it over seriously with some of the guys.

"Stan," I said, "I've been giving some strong thought to Kermac. What do you think?"

"Well, they're still on that contract and makin' double what we're gettin'. Everybody I talk to, though, says the place is pretty bad. Wet. Real wet."

"How wet?"

"Hauled out a man with the bends last week," he said with a grin.

"Very funny. What are you going to do? Are you doing to stay here or what?"

"A little longer, maybe. I got an application over in Jeffrey City now."

"Contract?"

"Yeah. Talked to some guys over there and they said they were doin' pretty good. Probably I'll just hold off a couple of weeks and see what happens." Stan, with his wife and their five kids, had come from Arizona and hired on when I did. We had been at Exxon about four months.

The next day I took off for "personal business," the catch-all excuse, and stopped over in the Kerr-McGee offices. A heavy-set man in an office much less pretentious than Exxon's looked up.

"What can I do for you?" he asked with a smile.

"Looking for work in the underground."

"Got any experience?"

"Two years. I'm over at Exxon now."

"Fill this out," he said, handing me an application.

I filled out the abbreviated form, listing only the mining experience I had. In ten miutes I was finished and handed it back.

"Okay," he said, after he had glanced over it, "can you take a physical this afternoon?"

It was the most perfunctory physical examination I had ever taken. This doctor—I assume he was a doctor—couldn't have cared less about what he was doing. He received his fee regardless of whether he took the time for a decent, conscientious examination, and I doubt whether I was in there ten minutes. Urinate in a bottle, take your clothes off, two arms, two legs, good, put your clothes on and get out of here.

The next day I brought the results of the examination to Kerr-McGee. The heavy-set guy behind the desk glanced at it and asked simply, "Start tonight?" There was little similarity between Kermac and Exxon.

I drove back over to Exxon to clear out my gear and give them advance notice that I had already quit. The reactions were about what I expected. Wiley Brooks said that he didn't blame me and wished me the best of luck. He also said that Exxon might have to reconsider its no-rehire policy. When I walked into the personnel office, I was no longer Mistah Voynick, just Voynick. The personnel manager acted as though I had personally betrayed him, and seemed incredulous at the fact that I would jump Exxon to go over to Kerr-McGee simply to get twice the money. He was just getting to know hardrock miners.

Tramping over to Kermac was a homecoming of sorts, since so many of the miners were from Harrison and Exxon. A lot of the guys wanted to know what took me so long. I told them it was company loyalty. We geared up in the dry and headed toward the red skeletal steel headframe. One of the miners pointed to my wet suit and said, "Hey, Pard, you better button that thing up and get that hood over your head."

The crew of eight miners stood on the cage, completely covered with yellow rubber except for their faces. As the cage abruptly started its descent, the first drops of water were carried by the cold draft into our faces. By the time we reached the pump station, thirty feet above the main level, the entire shaft was nothing but a torrent of ice water. My God, I thought, I bet they did haul a guy out of here with the bends. The crew shuffled off the cage, heads downward like a bunch of monks in saffron robes, and switched over to a new bank of high-capacity electric pumps which would handle the water for the shift. Those pumps removed 2,200 gallons of water from the mine every minute of every day, and the mine was only in its initial stages of development, with less than a half mile of drift off the station. This was going to be a wet mine.

Kermac was contract, so it was break your back all the way. Don't walk when you can run, and don't even pretend about safety. An engineer doubled as the official company safety man when they had to make a front. Kermac was drill, blast, muck, stand steel, and start over again. The contract was paid as a bonus based on the footage of developed drift on top of the straight $7.40 per hour. The water problem slowed things up a little; miners always had to fool with the pumps, and the mud in the drifts made it hard to move equipment and sometimes even to walk. There was no time for breaks down here. No one wandered around to socialize under the pretense of looking for tools, and shifters often worked right along with the miners, since they got a good piece of the contract too. Drill holes would be loaded before the steel was out of the last one, and the mucking would start before the vent system had a chance to clear the smoke and fumes.

At lunch the entire crew would come to the surface. Everyone was wet to a greater or lesser degree, and the bitter cold on the surface would freeze our wet suits and beards before we had a chance to plow through the snow drifts to get to the dry. Kermac at least provided a lunch table and benches for the crew's lunch time. The table was alongside the bulletin board, upon which were posted the accident reports from the Wyoming Bureau of Mines, good, detailed, graphic accounts of how miners all over the state had gotten killed or injured. Good lunchroom material.

"Well, how do you like it?" they asked me.

I'd answer with a noncommittal shrug.

"Pretty wet, huh?" with a knowing grin.

"That's the way I like it," I'd lie.

"Well, give it another week of crawling around in that ice water and tell me you like it."

They were right. After only one week Kermac got old. The cold and dampness would go right through you for eight long hours. Dress heavily and you'd drench yourself with sweat when you were working; take the clothes off and you'd freeze. It didn't seem nearly that bad when we got our first checks, though. Some good footage had gotten the hourly rate up to nearly $15 per hour, and, with a little overtime, the two-week gross would be around $1,400. Grin and bear it, I told myself, that's exactly what you're here for.

That rate didn't hold up every check, for there was always something to go wrong. The first accident that took the whole crew off contract for a week happened in the shaft. The muck was hauled out of the mine in two muck skips, steel muck containers, that were raised and lowered on cables directly adjacent to the main cage in the shaft.

No one really knew what happened, but the hoistman managed to raise one of the skips beyond the surface dump point on the headframe and right into the sheave wheel, the wheel on top of the headframe over which the 1¼-inch cable travels. The hoistman swore up and down that he was mechanically unable to stop the hoist, but no mechanical problems were immediately found, which led most of the crew to believe that he was either sleeping or not paying attention to the indicator drum in the hoistroom. At any rate, the cable parted and the seven-ton steel skip took off down the shaft like a bolt of lightning, roaring like a freight train past the main level station eight hundred feet below. When it smacked into the sump, the skip exploded into bits and pieces like a bomb going off. It was very fortunate that the skip didn't tumble in its guides or it probably would have reamed out the entire shaft, taking the main cage, the air pressure pipes, the sump pipes that carried the 2,200 gallons of water a minute, the other skip, and the high-voltage power cables with it. That would have been a charming situation for the crew on the bottom. If there had been anyone near the sump at the time of impact, he would've been the only man without a thing to worry about.

The job of getting the remains of the skip out of the sump was horrible. The twisted pieces had to be cut apart with torches while the men stood in knee-deep muck and ice water. Each piece had to be lifted individually on a winch cable to the main level. The men worked in a torrential downpour of ice water that must have been two hundred gallons of water per minute. There was some talk about chaining the hoistman to the other skip and repeating the act.

Although Kermac was a long way from production, some of the drifts were driven through the fringes of the ore body, and significant levels of radon were being recorded in the mine air. Rather than run vent tubing into those drifts to lower the levels, as required by law, management merely warned the miners verbally not to spend much time in them. Not much of a warning, just, "Hey, Pard, don't spend no more time in there than you have to. Got a lot of radon back there." No ropes, no signs, no vent. Just, "Hey, Pard."

There was only one thing that would change all that. Every once in a while we would hear it at the start of a shift. "Okay, you guys, Bureau of Mines is gonna inspect down here tomorrow, so we gotta get this place straightened up." It was as though God himself had ordered it; hundreds of feet of vent bag, all brand new, would be sent down, headings would be shut down, walkways repaired, signs hung, fire extinguishers replaced, and the piles of junk picked up. Readings would be taken of air flow and radon levels, and, at the end of the shift, the underground would be right up to standards. The safety man would take a reading of the radon levels in the bad drift, walk out with a satisfied smirk on his face and say proudly, "Got 'er right down there. We'll be ready for 'em." I used to love that. Who the hell are we supposed to be taking care of, ourselves or the Bureau of Mines?

One of the sad things in mining is the web of politics that governs and effectively limits the authority of the State Bureaus of Mines, the agencies that are primarily responsible for the enforcement of safety regulations. Their responsibilities are certainly not to the companies, but to the working miners who have no one else to rely on for the enforcement of regulations and for the creation and maintenance of safe and reasonably decent conditions. Any State Bureau of Mines inspector has the lawful authority to walk into any mine unannounced, grab the super by the neck, and drag him underground to witness a spur-of-the-moment inspection. Between high radon levels, high dust levels, below-minimum ventilation, and any number of other safety regulation deficiencies, an inspector could find, in half the mines in the state at any one time, enough violations to shut down part or all of the mine. But this never seems to happen. Whenever the mine inspectors pull their required inspections, there is always more than enough time to run around to correct the deficiencies.

The states are very cozy with the companies, and it is little wonder considering the direct employment, the indirect employment, the mineral depletion taxes, and all the general economic stimulation associated with the development and operation of a mine. What other choices do the states have but to go easy on the companies and make the whole package as attractive as possible to encourage the development of idle ore bodies? The Bureau of Mines, as the bureaucratic go-between, may just as well put themselves in the pockets of the companies and insure that everyone will be happy. The only people to get neglected are the little guys the Bureau of Mines was created to protect, the miners themselves.

In many mines, the miners know that to refuse an order, unless it is a monumental breach of safety regulations, will possibly result in their getting fired, or will at least earn them a reputation as trouble makers. Refusals may also guarantee a job in the sump for eight hours a day. A good example of that happened the day we came in to find a partial misfire, courtesy of the asses on the other shift. There was no air vent at all in the heading, a hundred-foot crosscut off the main drift, and the air hung heavy with a thick gray pallor of dust, smoke, dynamite, and nitrous fumes. The face was a mess, with live primers lying all over the place and cracked rock and slabs just begging to come down. The air was bad enough to irritate the throat and bring a stream of tears out of the eyes.

"We gonna hang some vent in here first?" I asked mechanically, fully expecting an affirmative answer.

The shift boss, usually a reasonable man, went through the roof. "Hell no!" he screamed, "we ain't got time for that, we're gonna shoot the sonuvabitch!"

It was obvious that he had gotten the law laid down to him by the super before the shift. The super here was a robot with all the personality of an engine block; his word was law and came down on a stone tablet. In my best dreams I could envision him chained up in the sump. I looked at my partners and in turn we all looked at the half-shot face, the miserable air, the slabs, and the live primers. Without saying a word we went in, rewired the mess, and shot it again. I wasn't quite ready to tramp, not just yet, not as long as the contract held up. Later, the shift boss calmed down and mumbled some kind of an apology for his spontaneous outburst, giving us an excuse having to do with his old lady.

Lunches, even though they were limited to a half hour, were a pleasant respite from the icy dampness of the underground. The guys at Kermac were all hard-core contract miners, and more than half were tramps. The conversation would rarely wander from the standard subjects of the contract and who was hiring where, all together with the well-worn repertoire of mine stories.

"Well, what did you guys get done so far?" was the standard opening. It was a reciprocal question of general interest, since the contract footage was figured on the total of the output of all the headings.

"We'll get 'er shot 'fore we go off, even with stone face lookin' over our shoulders."

"He down there with you? We never even saw him."

"You can have him last half of the shift. The sonuvabitch already aggravated my ulcers to bleedin' for four hours now."

The ritual of opening the lunchpails was performed in silence, the coffee poured, and the waxpaper rustled curiously to see what the old lady put in this time.

"Well, United Nuke is goin' to get started on their shaft as soon as we get a little warm weather. Gearin' up for it now."

"Might be good."

"Gonna be better than here, and I'll tell you why. Don't think for a minute this contract is gonna hold up. We been doin' too good for too long. They're gonna start screwin' around with it, you wait and see."

Another pause as everyone contemplates whether Kerr-McGee will refigure the footage and the bonuses they have been paying. The contract is the only reason half the hands are here.

"They cut it back, I'm gone."

"Where?"

"Back to Exxon for all I care. It's halfway dry and all you have to do is sit on your ass for a day's pay."

"They ain't gonna hire you back."

"That's bullshit. They'll rehire, got to."

"Think you could make it on a day's pay?"

"Know I could. It's my old lady that couldn't."

"Why not wait for United Nuke?"

"Goddamn animal outfit."

"This ain't?"

Quiet laughing. The rustle of waxpaper.

"What the hell is that?"

"You eatin' it?"

"Hell, no."

"Then don't worry about it."

"You know that copper hole's gonna get goin' over in Utah pretty soon. Anaconda or Kennecott, whoever it is. Kennecott, I think. They're takin' applications now. Supposed to be a big mine, five levels from what I hear."

"Where is it?"

"South of Salt Lake somewhere, probably not far from Bingham."

"Where you gettin' these applications?"

"Talk to stone face, he'll give you one." Facetious advice.

"I wouldn't ask that sonuvabitch for a drink if I was dyin' of thirst."

"That's good, 'cause I doubt if he'd give you one."

"That's the damn truth."

The eight miners sit at the table in the corner of the large dry, speaking in very low and slow voices that barely carry over the hum of the overhead gas heater. Curiously enough, no one looks up when he is speaking, but continues to peer into his open lunchpail as though there were something in it very engrossing, maybe a TV screen or a crystal ball. No one ever seems to interrupt the speaker, and no two people try to speak at one time. Lines are thrown out as if on a silent cue. The miner on the end is thin and is wearing a mangy three-day growth. He is only thirty-seven years old but looks fifty. Muttering a loud oath, he viciously packs an entire sandwich into a ball and flings it at the garbage can. The soggy ball hits with a dull thud.

"S'matter?"

"Baloney. All the goddamn money I give 'er and she's gotta buy goddamn baloney."

"Maybe she's tryin' to save some money?"

"Yeah, she's got to spend something on her boyfriend, don't she?" another voice cheerfully chimes in.

"You know, I'm thinkin' of goin' back to Arizona."

"Ain't much doin' down there now with copper, price got drove down too far."

"It'll pick up, always does. I'll tell you, this goddamn cold and damp is gettin' to me. Some mornings I get up and can't even move my hands, move my fingers, nothin'." A solemn murmur of agreement. "I've had about all of these cold, wet holes I can take. Man can make it on a day's pay in Arizona till it picks up."

Everyone sits quietly and listens to the bitter Wyoming wind rattling the aluminum on the sides of the dry. A two-foot-long drift of fine powder snow a few inches high is building up inside the dry by the door.

"Look at that damn snow in here. None of that down in Arizona now."

"You know, one of these days I'm gonna have my own mine."

"Yeah? Where you gonna put it?"

"Hawaii."

"What in the hell you gonna mine out there?"

"Anything. Pineapples for all I care. And hire a bunch of them naked titty broads to work for me."

"Jeffrey City's hirin' anybody that comes down the road now."

"What's that have to do with pineapple mines and naked titty broads?"

"Good money, that's what."

"You want to live out there? Christ, it's a hundred miles from the nearest phone pole."

"Did I ever tell you guys what we used to do out in Jeffrey? Me and Lummie, Lummie Atkins. You know him, don't you?"

"Tall? Thin? Used to work for Harrison a year ago?"

"Yeah, that's him. Looks like he ain't ate in a month, skinny. Well, we was partners over there contractin' out on a stope. The trick wasn't in gettin' the ore out yourself, it was in gettin' a goddamn train under your drawhole to haul it out. They couldn't run them trains fast enough. Hell you could wait all day for a motorman to get to you, so Lummie and me paid one of them. Old Jimmy Frye, gave him some money and some whisky and told him there was more where it come from. Christ, we were the only drawhole he'd stop under. It got embarrassin' the tonnage we were haulin' out of there. Makin' damn good money.

"Nother time they was payin' on the concentrate they got from your ore, so we got old Jimmy to take his train out by the mill and load up a car with rough concentrate, already milled, y'know, then he come back and we filled up the rest of the train with waste rock we had to get rid of. We didn't figure the percentage right or something, 'cause when they run it through the mill it come out so goddamn high they had a geologist runnin' around the next day tryin' to figure out what kind of ore we ran into."

Murmurs of approval. Every contract miner loves to hear a plausible story about how the company was beaten on its own contract. There is no such thing as a gift contract, and the companies will not offer contracts unless they figure to come out on top. Modern contract mining has evolved into the ultimate underground sport, the big game with the company where the prizes can double your hourly pay. The secondary reward is the simple satisfaction of knowing you bested the company at its own game. Stories like that foster the optimism that maybe it can be done again. And one story leads to another.

"Reminds me of the time I was diamond drillin' down in Arizona, Magma. My partner was this little squirrelly runt, small, but a goddamn good miner. Well, the two of us are working back in this dead drift gettin' footage pay for diamond drillin'. I don't know how they figured it out, messed up somewhere, but the two of us are really rakin' in the money. Hell, we had too damn much footage to turn in, y'know? If you give 'em that much, they're only gonna drop you down. So what we did was take all them extra cores and hide 'em in the ribs. Built us a bank, sort of. Little while later they took us out of there and put us in a regular heading. So for a month all we'd do is go back and turn in some cores every night. Paid us contract on that for damn near a month."

"Wish they'd come up with something like that here."

"I'm tellin' you, they're gonna be cuttin' back on that footage pay here."

"You know, when I was workin' down at that fluorspar mine in Walden, well—"

"Where?"

"Walden. Just across the line in Colorado. Well, we were drivin' this raise in really ratty ground, bad rock, some of the worst I ever seen. It was so bad that the engineer who was supposed to be measurin' footage was afraid to go up in there. Every day he'd ask us how much we got. If we got a four-foot round of it, we'd tell him six, then pretty soon we was tellin' him eight. All the time tellin' how bad that goddamn rock was to make sure he wouldn't go up in there, the bastard, and check for himself. We made some good money on that

A seasoned miner operating a drill. Each drill comes complete with deafening roar and brutal vibration. (ASARCO Incorporated)

one. They weren't sure but they thought they paid out a lot of money on that, too much. So they measured what they paid on paper and their goddamn raise come out twenty-eight feet above the surface."

Loud laughter.

"They want their money back?"

"I don't know. I got the hell out of there and tramped over to Harrison on the Exxon shaft."

The half hour is done. This is a contract mine and the lunchpails are closed and snapped shut.

I stopped at the bulletin board before suiting up. A new Wyoming Bureau of Mines accident report had been posted. Some guy in Green River was taking some air pipes apart and had a bad one. Either he had neglected to turn the air pressure off and to bleed the line, or someone else had turned it back on unknowingly. He loosened the bolts on a clamp holding a valve fitting, and when it came free, the valve blew through his head and killed him. There was still one hundred pounds of air in the line. Nice.

The crew put on the partially dry rubber wet suits and headed out in a tight group through the drifting, blowing fine powder snow. In the 4:00 a.m. darkness, the eerily illuminated headframe stood out as a beacon against the falling snow. A voice from within one of the yellow hoods asked the Hawaiian miner, "Hey, you gonna hire me for your Hawaiian mine?"

"Yeah, I might. Want to help sink the shaft?"

"No, I want to shaft the naked titty broads." The voice was nearly drowned out in the shriek of the wind through the icy beams of the headframe. The temperature was about five below zero.

True to expectations, management soon cut the contract back, and the once-heavy checks were suddenly lighter, but still heavy enough to prevent a mass tramp-out. Perhaps Kermac believed it could get the same amount of work out of half the bonus they had been paying. It wasn't long before any real enthusiasm on the crews faded away, even though a contract was technically in effect. No one was particularly happy, and more than ever, the talk turned to other mines.

Bob Hunter, one of the first Exxon hands I knew who tramped to Kerr-McGee, had a run-in with an old, grizzled, white-haired shifter who had come up from Arizona. No one had the faintest idea why he was hired, since we now had damn near two shifters for every miner, it seemed. The place was going to hell.

Bob and I were completing the form work in one of the new pump stations on the main level. Most of the large plywood sections were already in place, and all that remained was chocking the holes and seams with scrap wood and wedges so the cement, when it was pumped in, wouldn't pour all over the place as it usually did. We were out of the way in a dry place without the benefit of a shifter, and it looked like an easy shift. Then the light came bouncing down the drift.

"Okay, fellas," the grizzled shifter said, "I'm gonna show you how to chock this up right."

"Who the hell is he?" Bob asked quietly.

I shrugged. "I just hope he doesn't plan on staying here."

The old man pushed his way past me onto the narrow scaffolding, cornering Bob back by the forms. "Okay, fella, you just do what I tell you to." He made himself comfortable directly in front of where Bob was chocking seams. Thank God, I thought, there's only room for two people back there. I felt sorry for Bob.

"Okay, fella, take this here wedge and put it in right in there," he told Bob, gesturing with a wedge to an obvious hole in the seam. This went on for quite a while and I knew Bob, who was normally quite persevering and mild, was getting monumentally agitated. I went to the station to get away from it and bring back some more wedges. When I got back, Bob, his face set in a grim line, asked me quietly, "What the hell are they doin', makin' this place a retirement home for senile shifters?" I could afford to laugh, since I wasn't in there with the old man.

"Let's go fella, take this here wedge and shove 'er right in there."

The darkening look on Bob's face made it plain where he really wanted to shove the wedge. Instead, his quiet personality prevailed, and he merely walked slowly off the scaffolding without a word, went to the station, rang for the cage, and rode it up to the surface. In the dry he showered, changed, walked into the office to tell them figuratively where they could put the wedge, and drove home. I wandered around for a while, finally deciding I should put in an appearance at the pump station to see what was going on. This was at least forty-five minutes later, and the old shifter was still sitting there holding a wedge. When he saw my light, he asked, "Hey, fella, what happened to that other fella?"

"I don't know," I answered, "want me to go look for him?" I knew very well Bob was already halfway to Douglas.

"Yeah, maybe you better, fella. And when you find him, bring him right back here. We got a lotta work to do."

I wandered around for a half hour, then came back. "Couldn't find him," I said. "Looked all over."

He mumbled something I couldn't understand, then said, "You wait here till I get back." I waited until nearly the end of the shift, then drifted towards the cage with the rest of the crew. Bob was right, the mine was turning into a retirement home for senile shifters.

Bob's wife was about eight months pregnant, and it would have been a real strain to pay the maternity bills without the company insurance. Two days later he walked into the dry with a weak, resigned smile, and changed into his gear again. "Is he still down there?" he asked.

"Still down there." I answered. He just shook his head.

The development of the mine began to run into some serious delays. No longer could you count on coming in and working a heading. The first big project was to retimber the main drift that ran past the shaft station. The twelve-foot-square drift, when it was first driven, had been supported with standard square-set timber. But in only four months, the ground constriction had been so severe that the base of all the posts had been uniformly pushed in about a foot toward the center of the drift. Unless the caps, with all the weight of the back of them, are resting firmly and squarely on the posts, there is a good chance of eventually losing the drift. To counteract this, the ribs were relieved of muck, the posts were jacked back out to their proper positions, the floor of the drift was dug out and a bracing foot stull installed in each set. In theory, the four-sided sets could resist a force from any vector, and would thus stabilize the ground. It was a slow job, taking a couple of weeks. When we finished the drift it at least looked right, but not for long.

The ground continued to work, creaking and groaning as it pressed relentlessly inward. When the crews came back after a two-day shutdown, we could only look at what had happened and wonder just how much force the constricting ground actually had, how much the earth wanted to close the man-made artificial wound that was the drift. Every one of the stulls, and they were eight-inch timbers, had broken in the middle and forced itself upward through the floor of the drift to form a very uniform series of shattered wooden triangles, each triangle capped by the bright yellow of freshly splintered wood. It would have been something to have been down there when they started coming up.

New project now. Relieve the ribs again, get rid of the shattered timber stulls, and replace everything with steel. No headings had been worked now in nearly three weeks. I wondered what the boys sitting at the polished oak conference tables in Kerr-McGee's Oklahoma City headquarters were thinking.

Another project, this time to relieve some of the rock pressure by longholing into the ore body to drain some of the tremendous amount of water in the porous sandstone. A longhole machine, a rail-mounted pneumatic drill with drill-string clamps to allow the insertion of additional sections of steel as the holes deepened, was lowered into the underground and set up. The drilling began with several 1½-inch holes being drilled about 120 feet upward and laterally into the ore body. When the holes were completed, the steel string was withdrawn, and if the holes remained clear and didn't plug, a frightful column of water would blast out of each. The jet of clear, icy water shot out so forcefully that it would be impossible for a man to hold his hand against it. Someone estimated that each hole made somewhere around 150 gallons of water every minute.

Drilling on the longhole machine was murder. Nothing but noise, grease, and icy water. It didn't matter how you dressed, nothing would keep the drillers dry, particularly on the overhead holes where the crew worked in a frigid waterfall. All the time the two or three miners were working like mules and drowning like rats, two shifters, a foreman, and the super would stand there and watch, as though their mere presence would somehow be constructive. I knew it would only be a matter of time before I got the nod to hit the longhole machine.

By March, the steady snowfall had ended and much of the ground was already bare. Although the prairie grass and sage showed no greening this early, the wind, which only a few weeks before had been bitingly cold, had become almost pleasant in the afternoons. On one memorable day shift, enough people were off to cause some work reassignments and I stayed on the surface to work as toplander, loading the necessary supplies on the cage for the guys in the hole. Easy job, drive around the yard and rustle timber, put it on the cage, and ring it down. The afternoon was particularly mild, near sixty degrees. I drove out to the farthest reaches of the yard a half mile or more from the headframe, took off my coat, and stretched out on a pile of sun-warmed timber. Over the flatness of the prairie, the expanse of blue sky, dotted only with a few feathery clouds, was awesome. I wondered whether everyone noticed the sky that way, because of the plains and the absence of trees and buildings to obscure it, or whether one had to work underground in the dark, constricting drifts to appreciate it as I was then. The breeze was so gentle, the air so clean and everything so quiet. No oil and water mist spewing out from thundering drills, no smoke and fumes from the dynamite, and no cold, hard rock to fall out of that robin's-egg-blue sky. It was difficult to look at the prairie on that beautiful spring day and try to visualize the tiny drifts 850 feet below that had only one way out, where men just like me were sweating and cursing in the darkness at that very moment.

A group of antelope I had been watching drifted nearer and nearer. How any creature so gentle and graceful, so frail looking, could survive the high plains winter was a miracle. Antelope are perhaps the most beautiful of any animal, and against the drabness of the prairie their delicate reddish-brown color, white rumps, and soft white lines across the throat appeared brilliant. Off to the side, keeping his distance, was a buck with massive, jet black horns. The king of the hill. A group of jackrabbits, the fattest I had ever seen, with magnificent, soft, rich gray fur, was busy chewing on something on the hillside. Instinctively, I wished I had a twenty-two, but it was only a brief, fleeting thought. In the mood I was in, I couldn't have shot anything anyway. Not even the super.

The thought of the super and his stone face made me think it was about time I put in an appearance somewhere. I had been lying on that pile of timber for two hours. I drove the pickup back toward the distant headframe, almost hoping somebody would say something to me. When the hoistman saw me getting out of the truck, he ran over.

"Where you been?" he asked excitedly. "Your shifter's been goin' crazy tryin' to get you on the phone."

"Good." I knew I wasn't long for Kerr-McGee or, for that matter, any mine.

The next day I skipped the car pool and drove in alone so I could stop in Douglas and have a beer with Stan on the way home. I pulled up to his trailer in the late afternoon, saw his truck wasn't there, but knocked anyway. The manager came walking over and asked if I wanted to rent it.

"Just looking for Stan Griffith," I said.

"You'll have to look a lot further than here. He's gone. Left yesterday."

"Where?"

"Jeffrey City. Took the whole family with him and cleared the trailer out. Said he was gone for good."

I went over to the Saddle Rock for a beer. The Saddle Rock was usually quiet in the afternoon, but could provide some real front-line excitement and entertainment at night when the cowboys, miners, and roughnecks were all loaded. I was in there once, sitting at the bar with some of the crew, when the bartender suddenly collapsed and crashed to the floor. Everyone sat there calmly sipping his beer as though the poor guy were doing nothing more than tying his shoelaces. Finally, a deep voice boomed out from the end, "Well this is one piss-poor way to run a bar. What the hell does a man do when he wants another beer?" Finally some cowboy got up and walked behind the bar to see what was keeping the bartender. The same deep voice was raised again. "Hey, Pard, as long as you're back there, mind gettin' me another one?" It all goes to prove that if you look at enough violence and injury, you'll get inured to it.

This afternoon there were a couple of Exxon hands sitting there. I parked alongside them on a stool. "What're you guys doing over there?" I asked.

"Same thing you're doin' over at Kermac. Nothin'."

"What's this I hear about Stan?"

"Tramped out yesterday. Takin' some contract in Jeffrey City. You hear what happened today?"

"You guys got a round out for a change?"

"Very funny. You know Harry Linton?"

"I know him by sight, never really talked to him," I said.

"Got hurt bad."

"How bad?"

"Got himself caught between the motor he was runnin' and some steel. Broke damn near everything on 'em. Worst mess I ever seen. They took him here first, then they had to take him to Casper and his heart stopped on the way. They got him goin' again and I guess he's still alive now, I don't know. He ain't goin' be worth anything if he comes out of it."

We sat there a while, no one saying anything. One of the Exxon hands bought another round. "Got your income tax done?" he asked.

"Yeah," I said, "should have a good chunk coming back."

"How many dependents you declare?"

"None," I said, "I like to make sure I get something back."

"That ain't the way I do it," he said, "I declare more than I got. You know why? Look at what happened today. If I was to get killed, I'd feel pretty goddamn rotten knowing the government had some money that belonged to me."

I pondered that morose bit of logic for a while, finished the beer, said good-bye, and drove to Casper. When I got back home I took a look at my bankbook and checking account and decided there was no sense in being greedy. It would be the first shift I didn't like.

I didn't have long to wait. When I walked into the dry the next day, my shifter said, "Pard, you're gonna be longholin' with Jerry today." I nodded, put on all the rubber I could beg, borrow, or steal, snapped all the snaps, and resigned myself to the inevitable fate of the longhole machine.

On the way to the headframe, Jerry said, "Pard, you're gonna love it."

I loved it all right. At the end of the first hour I was soaked, frozen, and my hands were numb. The water cascaded out of the overhead holes in sheets, and on the other side of the waterfall I could see the lights of the audience, a shifter, a foreman, and stone face himself.

I yelled above the roar of the drill and splashing of the icy water, "Jerry, how the hell can you do this everyday?"

He managed a wet smile, then yelled back, "Won't be doin' it forever. One more year is all, that'll make ten, and then . . ." He gestured up with his thumb.

"What are you going to do . . .?" I asked, gesturing up with my cold thumb.

"Anything. Build houses maybe. Work with my old man. But only one more year down here. You like longholin', eh?"

"Yeah, this is great. Sorry I missed it all till today."

He grinned at the sarcasm. I grinned back, with the icy water running off me in torrents. The two of us stood there in the deluge grinning like a pair of fools. The audience did not grin.

At lunch I showered and left my soaked clothes lying in a soggy heap in front of my basket. Someone suggested I hang them up or they wouldn't dry, assuming I had forgotten about them.

"That's okay. I'm not going to need them after lunch anyway," I said.

It was a quiet lunch. There were a few comments about the guy from Exxon who was still hanging on in a Casper hospital. Then some lighter ones about the nice weather and how it was nearly time to start fishing.

When lunch was over, I exchanged some grins with the crew and that was it. They suited up and headed out to the headframe for another ride down in the cage, while I started changing into my street clothes. After a few minutes stone face crashed through the door and asked brusquely, "What's the matter with you? Cold and wet? So's everybody else, let's go."

"No," I told him in a very nice voice, "actually I'm warm and dry, and that's the way I plan to stay. If you're in such a rush to drill your goddamn holes, you can drill 'em yourself."

"What?" he asked, this time in a much milder voice.

"I just quit," I told him with the same silly grin I had been wearing for several hours.

"Quit?" he asked, as though he didn't understand what the term meant.

"Quit," I repeated. "Quit. Q-u-i-t. Do you know what quit means? It means I don't work here anymore."

"Oh," he said in an enlightened tone of voice, not as though he understood that I just had severed my employee relationship with Kerr-McGee, but rather as though he had just learned the meaning of a new word. With that he turned and left.

Since I didn't drive my own car that shift, I stood out on the dirt road that stretched across the prairie and hitchhiked. In a few minutes I got a lift from some kid living in Casper. He had just put in his application to work underground as a miner's helper at Kermac. Understandably, in the presence of a genuine miner, he was very talkative.

"What's it really like in the mine?" he asked, almost giggling with childlike anticipation.

"Oh, it's nice," I said in a monotone. "You'll love it."

This was the second warm spring day in a row, warm enough to ride with the windows down. I felt like the weight of the world was off my shoulders.

"But why did you quit?" he asked at length, seeing no logical reason in light of the pay and how wonderful everyone told him the underground was.

I smiled to myself, thinking back to that January day at Climax, Colorado, many years ago when old Jim Wizen had gotten knocked on his ass by the first slab I had ever seen. His immortal words came back clear and true, their worth and impact undiminished by time. *Gettin' too old for this shit.*

CHAPTER VI

AND SEEK NOT YOUR FORTUNE

Several years ago, on a hot summer day, I passed through Jerome, Arizona, on one of those many trips that seem to have started nowhere and led to nowhere. Not that it was the beginning or the end that really mattered, but rather the journey itself that was the important thing. The sign on Jerome's deserted main street said "Paul and Jerry's Saloon," and that was good enough for me. I stopped in to rinse the Arizona dust out of my throat, and struck up a conversation with the old man behind the bar. It was an old bar in an old building, with an ancient player piano, dusty chandeliers, and a collection of mementos from days gone by hanging on the walls. I slid off the stool and sauntered along the wall examining the exhibits that, for some reason, seemed of a personal nature. There were some old carbide miners' lamps and many faded, cracked photographs of groups of miners. I studied the photographs, looking closely at the gear the miners wore, thinking how different, yet how similar, things were today, and looking closely into the frozen, stoic expressions of the men wearing the lamps.

"They were taken a long time ago," the old bartender volunteered, leaning across the weathered, scratched bar. "Fifty years ago, back in the twenties, way before your time."

"Are you in any of them?" I asked, knowing somehow the answer would be yes.

"A couple," he answered, with a touch of pride in his voice. "But I knew all the men in 'em. Every one. That was all right here in Jerome, Consolidated Verde. All gone now."

I moved back to the bar stool and said, "Well, I can appreciate those pictures more than you think. I've done a little mining myself over the past few years." He looked up with renewed interest and asked where. I told him.

"Guess things have changed quite a lot in the mines since the days when I worked 'em," the old man reflected.

I thought about that for a while. The equipment had been improved in many ways. Drills were much lighter and more efficient, dynamites were greatly improved, air ventilation was much more efficient, new equipment had been introduced, virtually all safety standards had been upgraded, at least on paper, and the mines were statistically and indisputably safer for the working miners. But how much improvement had been made in rock bars, double jacks, mucksticks, and human backs and lungs? And how much improvement in the elimination of the slabs that still crashed down on human bodies? It seems that most of the heralded improvements in American mining over the past century have functioned to increase the operating efficiency of the companies, allowing them to mine lower grades of ores that were once deemed economically infeasible. ST-8s, for example, allow a single miner to move more muck than any ten miners could have dreamed of moving fifty years ago. While such machines also reduce the chance of injury by reducing the number of man-hours necessary to remove a given tonnage of muck from a heading, they simultaneously introduce new hazards that were not present before; the danger of enormously heavy equipment rolling through the narrow drifts, and the omnipresent fire hazard created by the hundreds of gallons of flammable petroleum products sloshing around in thin metal tanks. In the old days when underground ventilation was a haphazard affair at best, mine fires could often be contained. Today, with vastly more efficient vent systems powered by enormous fans moving great quantities of air through well-designed flow patterns, the deadly products of combustion are also distributed quickly and efficiently throughout the underground. This, ironically, is exactly what happened at the Sunshine.

Yet, while there has been much new equipment introduced into the underground, many of the techniques and basic procedures in use a century ago remain unchanged. "Rock Mechanics" is the title of a course taught to today's mining engineers. It supposedly imparts the knowledge necessary to, among other things, devise better ground-support techniques. But in countless mines throughout the West, the square-set timber supports are identical to the ones erected in 1880, and they are still installed by miners manually lifting the heavy timbers into place and securing them with ax-driven wedges. The sole difference is that the sets are larger, in accordance with the larger drifts, which makes possible the use of heavier equipment. Everything else, right down to the ax that drives the wedges, is the same. The innovations in underground mining have done far more good for the companies than for the miners themselves. In fact the lot of the hard-rock miner has remained basically unchanged, and his situation must rank as one of the most static in American labor.

I explained all this to the old man, who smiled and said, "That's the same thing we used to say when I was underground, how little it had all changed. But I thought now, since I'd been out so long, it'd be different."

I shook my head no.

"Well," he went on, "I guess mines will always be mines and miners will always be miners. You know, back in the twenties, at shift change, the miners would be five foot deep right here at this very bar." He gestured to the empty, worn wooden floor.

"That hasn't changed, either," I said with a grin.

"Still, though, even though it was rough, I can't complain. The mines were good to me. I made money, good money." He paused to laugh. "Well, good money then, maybe. Fifty cents an hour. And there was no place else I was going to make that back then. No place. Yessir, the mines were good to me."

I wondered whether it was the passing of the years that had allowed him to forget the sweating and cursing he did in those tight drifts, and to recall only the paychecks and the boisterous back-slapping when his partners were five deep at the bar. The reasons for going into the hole in the first place hadn't changed either. Very basically, it came down to simple economic need. Even today, the only reason a person goes into the hole is for the paycheck, or for a curious combination of local custom or tradition, the following in the footsteps of a father or an older brother. When a high school kid graduates in a mining community, and college isn't in the picture and family ties are strong enough to hold him, he can either work in the local drugstore for peanuts or hire on at the mine and immediately start earning three times as much. It is a strong inducement, since those mine paychecks will buy that hot car, pay for a lot of beer, and impress a lot of girls. Then, quite suddenly one day, the realization is there that you can no longer get out, and that all mining will prepare you for is another mine somewhere else.

Not every man who tries it can stay, even if he wants to, for it takes a unique psychological makeup to cope daily with the darkness, the danger, the confinement. Those who are able to cope soon develop a strong sense of pride in their confidence and their abilities, and in the knowledge that not every one could make it as an underground miner. A good number of miners, whether they choose to admit it or not, revel in that daily risk and danger, finding it fulfills a basic psychological need. These same men, sitting at a desk or running a drill press in some shop for eight hours, would leave with a gnawing sense of something missing, a feeling of unfulfillment, an emptiness. Mining, since the underground dangers constitute a very real and tangible enemy, fills a definite void in these people by giving them something to fight, to battle, and to beat. And every day they walk out of a mine in one piece, they enjoy the deep-seated satisfaction of having looked it in the eye and beaten it again. And the simple fact that they are going to risk it all again tomorrow produces a titillating sensation tantamount to a thrill and provides them with a constant anticipation that, oddly enough, tends to enrich their very lives, to make their families mean more, their beer taste better and their free time more valuable. The simple things in life are no longer quite taken for granted. Ask a hardrock miner about that and he will tell you you are crazy. I know, because I have. The miner will say only that he has the greatest respect for the dangers of the underground, even a controlled fear of them, and to think otherwise is a conceptual fabrication. Yet when the one-ton slabs come down and miss him by inches or crash into the exact place he was standing two seconds ago, the first thing he will do, nine times out of ten, after controlling the rush of adrenaline, is grin. A big, wide grin. "Goddamn, Pard, she almost nailed me." *Almost*. Not a grin to bolster confidence or suppress fear, but a grin because he beat it again. Whether it was because of agility, experience, dumb luck, or whatever, he gave it a fair shot and beat it, got away with it, took a chance and pulled a card to fill the inside straight. I've felt it myself and I've seen it on the faces of many, many miners.

I once read a paper written at a midwestern University in which a professor described these tendencies as normally present in a certain percentage of men. He said that during the American frontier days, there were more than enough constructive outlets for them. It was these men who, supposedly, were the mountain men, the railroaders, the cattle drovers, and the other more or less stereotyped characters who have become legends in American history. Today's society, in a nation that is no longer expanding and developing in the unrestricted sense of Manifest Destiny, has become overly tame, too organized and regulated to provide available constructive outlets for these men. The paper went on to hypothesize that many of our contemporary misfits, drifters, criminals, and winos are merely men who lack the proper outlets in a world that strives almost blindly for safety and security. Underground mining would certainly be one of the few remaining professions today that would offer such an outlet, providing enough inherent risk to satiate these men. The mines, in this respect, are generous indeed.

It is because of the dangers of mining that the bonds of friendship and trust that develop between partners or members of a crew are unusually strong. Underground mines are no place for the facades of artificiality that are so prevalent on more conventional jobs where a man's performance and reliability may be disguised or tempered by mountains of paperwork, large staffs, and numerous opportunities to use the efforts of others to make himself look good. In a mine, it is you and the rock, you and danger. Either you face it and do it, or, at the end of the shift, it is not done. The utter simplicity of two men's working together in a heading makes valid excuses few. Faking the work, as well as hiding a rampant fear of the underground, is impossible.

Another basic trait of underground miners is honesty. Any dishonest man would choose one of the many easier ways to make money before ever considering going in the hole. It is only natural to trust men like this, and that ultimate trust that is extended in the headings is easily carried over to the off hours, and it is upon that trust that extraordinary friendships are built.

Modern hardrock tramps, who are among the most independent and unusual men in the country, are a true vestige of the frontier. A good tramp miner, following the good contracts, will earn $30,000 per year. He can find a job in New Mexico, Arizona, Colorado, Utah, Wyoming, Idaho, or Montana any day of the year and he knows it. He is basically a restless person, one who can feel something closing in on him when he stays in one place too long, a person who thrives on and is refreshed by changes in climates, surroundings, customs, and people. I have seen miners tramp out for vague non-reasons like "it's time," "been here long enough," and "feel like it," as well as the more tangible reasons like higher contracts elsewhere or getting angry at the super. They enjoy an independence and freedom that is rapidly vanishing in the country. For many it is an endless march, simultaneously sad and happy, a compulsive journey to someplace different, yet to the same thing they have just left.

Company loyalty is non-existent, an unnecessary and binding allegiance, for the only thing better about any mine is the number imprinted in red on the paycheck, and that is subject to change. Although every mine is essentially the same, tramp miners are still perpetual optimists, believing deep down inside that the next mine is somehow, in some vague way, going to be better.

Some time after I had left the mines, Stan Griffith, whom I first met hiring on at Exxon in Wyoming, sent me a letter. When I received it, he had already tramped Jeffrey City and was contracting at another uranium mine near Thoreau, New Mexico. He had previously compiled a series of seven short stories, vignettes actually, of his experiences as a tramp miner. When he heard I was considering a book about the hardrock miners, he offered me the use of those stories.

Steve,

I could write you a long piece on bonus mining and how the companies I have worked for have ways of cheating men out of it and how they use it to make men work harder for nothing. I could also write a piece on what it takes for a woman to stay with a tramp miner and how these gals are a breed of their own too. If you need any help or have any ideas I can help with let me know. And Steve excuse my bad spelling, I'm just a tramp miner you know, not a college boy. These stories need to be rewrote as they should have been longer. Feel free to improve on them as you see fit.   Stan

I selected two of these, *The Tramp Miner* and *The Tramp Miner's Wife*, to enhance the picture of the modern tramp miner. These vignettes are, I am sure, a self-portrait of sorts, and had such honesty and simplicity that I saw no need for the improvements he suggested.

THE TRAMP MINER

A tramp miner is exactly what the two words imply. A tramp: one who moves about the country of his own free will. A miner: one who works underground in the removal of ores. Put them together and you have one hell of a breed of man. There are different types of miners, the home guard miner, who is a man who started in one place and stayed with it, the one-type miner, who has never done but one type of mining, be it stopes, raises, etc. But a tramp miner, now he's a different breed, a free soul, a sudden summer storm comes in and he's gone the same way. He's a natural born explorer. Whoever it was who put the title on his name, I toast him, and say to the unknown giver of names you have done him well and his title shall be carried for years to come. I thank you. But my wife doesn't. The tramp starts as all miners do, in a mine that was probably the best job he could get and once he gets into the thing called bonus or contract mining he is hooked and never leaves the trade until he is dusted so bad that he can't breathe, or like so many before him he's carried out and put to rest with his Maker.

To these I say God bless you, no one else will.

Contract mining is two things. It's first and most important extra money for extra work. Second it's a challenge to beat the next miner and show him how much you can make the company pay. Contract mining is usually why a man stays with this job which isn't rated as high as it should be since mining is one of the most dangerous jobs on earth. The enemy or company want to get as much ore out of their mine as fast and as cheap as they can. In order to do this they came up with bonus, if you pay a pair of miners 100 dollars a day in wages and they move one ton of rock which is valued at 100 dollars a ton then they break even, no profit, but if they tell them that if they move two tons a day for ten dollars extra they will do their hardest to break them three or four tons a day. Now these figures are fictitious but they will give the reader an idea of how bonus mining works. The only problem is that most big companies have a way of changing their systems just about the time a miner really has some good bonus money coming to him. In most cases he rebels the only way he knows how, he sits on his ass until the company gets nervous. In the case of tramp miners, it only takes him one time maybe two to get fed up and tell the company where to shove the drift they want drove. The home guard miner usually keeps on taking it because he does not want to leave that section of the country or is afraid he can't cut it someplace else. That's why the tramp is a breed of his own, if you treat him good he stays and does you good. But you cheat him and before you know it you need another miner to take his place.

THE TRAMP MINER'S WIFE

The middle aged woman stood over the boxes she was packing and muttered something under her breath she didn't want the five kids to hear. What did you say, Baby? the stocky man said as he stopped at the door holding a box of clothes in his hands. Not a damn thing you no good son of a bitch, just keep loading the truck before it gets dark. Smiling, the tramp miner just shrugged his shoulders and headed outside with the box of clothes to the truck.

It had been this way ever since they were married ten years ago when she met him at a dance in their home town, which happened to be a mining town, she thought in disgust. He was so good looking and strong before the fellow in Butte broke his nose with the cue stick or the one in Green River put the scar on his forehead with the broken beer bottle. As she thought of the times past she remembered their firstborn and how proud he was, so proud that he got drunk for four days and lost his job. But he was real good about it, he had the car packed for her so when she got out of the hospital they could leave right away.

Now they were off again to new adventures or so he thought but she knew it would be the same, and with any luck maybe they could make it a year this time. Ha, she knew better. Mommy! Mommy! the tugging at her dress snapped her out of her dreams of the past. I want a drink of water please. Go tell your big brother to get you one. Mommy is very busy right now Sweetheart. As she left the room she heard the icebox open and the tab from a beer can snap off. Her expert packer upper, mover outer was taking a rest break. She entered the kitchen. He looked up at her with that old look in his eyes and she knew what he was thinking. You want to mess around Old Lady? Go play with yourself Big Stud was her reply as she opened the icebox for a beer. What the hell's wrong with you all of a sudden? Now she wished she hadn't got so smart mouthed. You know what's wrong with me! I don't know why we have to move so much. Well Jesus Christ woman, this ain't the first time we moved, you know. And it won't be the last time either I'll bet! Best shut your mouth woman! Although he had never laid a hand on her she knew it was time to shut up so she did so. Will we be able to take the dog with us this time? You bet, Babe, everything goes but the kitchen sink. As she sat sipping her beer she wished now that he could have talked her into buying that trailer five years ago. At least they would have a home when they got there.

As nightfall was upon them and they headed north on the interstate she looked at him still full of life and realized that if he had to stop moving around it would kill him, take the adventure right out of his life. Besides, she wanted to see beyond the next headframe too, 'cause after all the packing was done and the kids asleep in the camper it was just him and her and the great love they had for each other. Deep down inside, she liked to tramp too.

I certainly hope she did, because the next time Stan tramped, which was only four months later, it was no ordinary day-or-two drive to the next mine. Stan, his wife Kay, and their five kids tramped to South America, to Tacna, Peru, where he managed to climb out of the hole on top of the Andes as a shift boss in the huge Southern Peru Copper Corporation open pit.

Many people, particularly in the East, upon hearing the term 'underground mining' think immediately of the Appalachian coal mines, and for good reason, since those mines have received tremendous attention in the news media. Almost every time one picks up the newspapers or turns on the evening news, there seems to be some coal-related development deemed newsworthy, usually something to do with labor problems, a wildcat strike, violence stemming therefrom, a union election, or, occasionally, a mine disaster.

Although it may surprise many, there are virtually no ties, socially or culturally, between the underground coal miners and the western hardrock miners, even though they wear the identical lamps and self-rescuers, and dodge the same body-breaking slabs. Hardrock miners, as a matter of fact, look down upon the coal miners for some reason, apparently feeling that the work is "too dirty" or "too dangerous," or that it simply wasn't worth it to "crawl around on your hands and knees under a four-foot back." Maybe, but the only advantage I could see was that a piece of heavier granite falling from a higher place could do the job a little cleaner. The break between the two might better be traced back to geological conditions that fostered different sets of social conditions. The vast coal fields of the West are mined primarily by open-pit methods, although there are underground coal operations in Utah and Colorado, and the surface mining is not mining in its classic underground sense, but rather a huge ground-moving operation. The coal fields of the East, the historical source of the nation's coal, occur in a manner which established permanence among the men who worked them. The coal deposits were extensive enough that twenty or thirty mines would work a particular field. Towns grew up around them, and, if a mine should shut down, the miners could simply seek work in another one nearby. There would never be the spectre of leaving home, of breaking the family ties, of seeking work five hundred miles away. The local roots strengthened, and permanence became a byword in the coal miners' lives. The mines became unionized, at least many of them, and the miners eventually chose to face life on their home turf, for better or worse. What happened then, over the course of long strikes or major declines in production or demand, with the depression, poverty, and resultant welfare problems which have become so publicized?

Hardrock miners have managed to avoid most of these blights, mainly because of the historical absence of geographical roots and their tendencies to tramp at the slightest production slowdown, contract reduction, or other reason, however whimsical. Their basic independence also accounts for the low-keyed and spotty presence of the unions in the western hardrock mines. The miners in many areas of the West simply prefer to take care of themselves rather than rely on a union. In the four mines in which I worked, only the Climax hands were represented by a union. Most of the hardrock mine towns, or those communities close enough to a mine to become the home for many of the workers, are vibrant, growing communities, many with boom economies. These towns, even upon closure of the mines, will never fall prey to the economic scourge which took much of Appalachia, for the miners will be gone before the mines close. Hardrock miners are one of the last remaining groups that honor the work ethics that built this country. If the pink layoff slips come out with the checks on

Friday, they won't be standing in the welfare lines on Monday. Most will already be picking up their brass in another mine.

Throughout the Rocky Mountain West there are about sixteen thousand hardrock miners, a relatively small work force considering the vital effect they have on the American economy. Yet for all their productivity and the uniqueness of the job, the hardrock miner does not have a degree, a license, or any professional recognition or formal rating. All he has is his back, his experience, his knowledge, and the psychological base that allows him to cope with the underground environment. The technicalities of his trade are passed on by word of mouth, for there are no books explaining how to run a jackleg, how to rockbolt, or how to timber bad ground without getting killed. And when miners change jobs, as they so often do, they walk into a new mine without so much as a single shred of paper to verify their experience and competence.

I remember well the words of the old miner in Jerome, Arizona. "Guess things have changed quite a lot in the mines since I worked in 'em."

Everything considered, they have changed very little.

The old man also said that the mines were good to him.

I guess, somehow, when all the complaining and the straining and sweating was done (and since I managed to walk out with all the flesh and bone that I walked in with) the mines were good to me, too. To this day, I don't know who got the better of the deal. They were always there when I needed them, no questions asked. And in them I learned what the real work ethic was, reduced to its simplest terms. I also realized how little of it is left in this country. But the greatest thing the mines gave me was the acquaintance of the finest people I have ever met, people who were willing to judge a person because he was a person, and who measured merit and sincerity and not the artificial trappings that play such a major role in so many avenues of American society today.

Even with their honesty and straightforwardness, they are still men who, when it came time to say good-bye, tried to conceal their emotions behind a mask of hard rock. It was always an awkward confrontation, a few clumsy handshakes, and "Well, Pard, when you come through again, you know the door is always open."

And they never did forget because, years later, true to their word, the doors *were* always open. In fact, sometimes their memories amazed me. Several years after I had tramped Climax, I stopped back to see my friends, including Ted Wiswell and my first partner, Dwayne Black. The three of us were sitting in Dwayne's place on a strikingly clear January day in Leadville. It was the first time I had seen them in four